W E I S E R   ⚥   C L A S S I C S

THE WEISER CLASSICS SERIES offers essential works
from renowned authors and spiritual teachers, foundational
texts, as well as introductory guides on an array of topics. The
series represents the full range of subjects and genres that have
been part of Weiser's over sixty-year-long publishing program—
from divination and magick to alchemy and occult philosophy.
Each volume in the series will whenever possible include new
material from its author or a contributor and other valuable
additions and will be printed and produced using acid-free
paper in a durable paperback binding.

LIBER NULL

AND

PSYCHONAUT

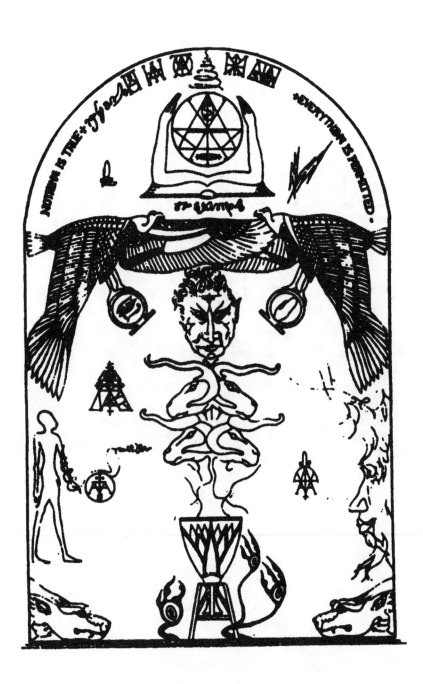

LIBER NULL

AND

PSYCHO NAUT

THE PRACTICE OF CHAOS MAGIC

Peter J. Carroll

FOREWORD BY RONALD HUTTON

WEISER
BOOKS

This edition first published in 2022 by Weiser Books, an imprint of
Red Wheel/Weiser, LLC
With offices at:
65 Parker Street, Suite 7
Newburyport, MA 01950
www.redwheelweiser.com

ISBN: 978-1-57863-766-9
Library of Congress Cataloging-in-Publication Data available upon request.

Series Editors
Mike Conlon, Production Director, Red Wheel/Weiser Books
Judika Illes, Editor-at-Large, Weiser Books
Peter Turner, Associate Publisher, Weiser Books
Series Design
Kathryn Sky-Peck, Creative Director, Red Wheel/Weiser

Illustrations in *Liber Null* by Andrew David
Illustrations in *Psychonaut* by Brian Ward
Typeset in Arno Pro

Printed in the United States of America
IBI
10 9 8 7 6 5 4 3 2 1

*To all those Psychonauts with whom I have stood in midnight forests,
in temples, in subterranean chambers, and atop mountains,
invoking the Mysteries . . .*

Acknowledgments

I wish to gratefully acknowledge all the people who, over the years, have helped make this book possible. To Ray Sherwin, who helped make the first version of *Liber Null* available in 1978, and who worked with me to produce the revised version of 1981; to Christopher Bray of the Sorcerer's Apprentice, who helped keep *Liber Null* in print and produced a limited edition of *Psychonaut,* and who helped by making these available to seekers through his bookstore; to Andrew David, who did the illustrations for *Liber Null;* and to Brian Ward, who did the illustrations for *Psychonaut.* My thanks to you all. The present edition is a completely updated and edited version of both works.

Author's Note

Liber Null and *Psychonaut* were written for serious occult seekers and therefore contain some powerful rituals. These rituals and exercises should be performed by readers who are in good health. If one suffers from heart disease, epilepsy, or any chronic disease, please do not use the material in this book. The author and the publisher will not accept any responsibility for misuse of this material, nor will they accept any responsibility for anything that may occur when readers use the exercises discussed here.

Contents

PSYCHONAUT

Foreword

It may be suggested that there have so far been three leading theorists of modern Western magic. The first was Alphonse Louis Constant, alias Eliphas Levi, in the mid-19th century, who established that in order to work magic, it is necessary first to change oneself and ultimately to understand the universe. The second was Aleister Crowley, in the early 20th century, who drew on both Eastern and Western traditions to provide conceptual frameworks within which both aims might be achieved. The third has been Peter Carroll, in the late 20th century, who placed the same aims within an ultra-modern, or post-modern, context of cosmology and morality.

Much of the nature of the achievement of each man has derived from his essential nature: Levi as a priest, Crowley as a poet and adventurer, and Pete as a scientist. Other great magicians, such as those of the Hermetic Order of the Golden Dawn and Austin Osman Spare, have contributed much in the way of practical techniques; but these three stand out as theoreticians.

The present book is the fourth edition of *Liber Null* to appear and the third of *Psychonaut*; and it is the second joint edition of them. The ideas contained within this new edition were therefore developed in the 1970s, honed in the 1980s, swept much of the Western magical world in the 1990s, and have remained influential since. Looking through the previous editions, it is remarkable how little they have changed in essence, which is a tribute to their raw original power and enduring strength. The only real alterations were early and minor. Like most magical traditions, the persisting ideas are a mixture of previous thought with brand-new concepts.

As the book itself makes clear, it draws directly on approaches to the cosmos—animism and shamanism—that are literally prehistoric. On the other hand, the place accorded to animism in human development is specifically that argued by late Victorian anthropologists like Sir Edward Tylor, while the view taken of shamanism is specifically that propounded by Mircea Eliade in the 1960s.

The ceremonial tools that are recommended to magicians have been used for thousands of years, and the ideas that magical workings should draw on an eclectic range of traditions in the cause of getting practical results, and that magicians are not fettered by normative social morality, are both found in the ancient Greek magical papyri. The Horned God, as the most important old pagan image of male deity, was originally developed by Margaret Murray in the 1930s, and the exact image by which he is portrayed in Pete's work is that created by Eliphas Levi and it is invoked in the ecstatic manner of Crowley. Classic Indian and Chinese techniques are recommended to obtain trance and an oblivion of desire, offered in conjunction with one developed by Austin Osman Spare.

All this, however, represents the stage that is set, and the props recommended, for the practice of magic in the book. The central ideas are the author's own and have an unmistakable context in the late 20th century. The most obvious of these is their relationship to the cutting-edge scientific thought of the period, especially chaos theory and quantum physics. More profound is the manner in which they represent in many respects the spiritual dimension of an anarchic individualism, in which the essential unit of society is the lone person, and the greatest goal of life is the full development and expression of that person's self and the realization of her or his potential.

All people are regarded as having an entitlement to act, desire, believe, and achieve whatever they wish as long as they do not encroach on the ability of others to do the same. To some extent, this morality is presaged in the precepts of Crowley and embodied in the famous ethical tag of modern pagan witchcraft, or Wicca, "Do as you will but harm none." It is not quite the same, however, because Peter Carroll's ethic does accord the right to its readers to harm others, not merely in self-defense but at any point at which any of them consider it to be justified.

Another very modern aspect of his system of magic is the complete absence of religion, even though divine forces and figures, including deity forms and spirits, abound. The distinction is that there is no injunction to worship any of those entities, and they are not expected to give laws to humans or supervise their conduct, and there is absolutely no room in the whole system for a priesthood, represented by magicians or not.

Instead, a cosmology is declared that allows for great cosmic powers while nonetheless giving human beings complete freedom of action. Those powers represent bodies or fields of natural energy, providing an essential bonding to the universe, on which humans can draw for their own needs. Out of them can be fashioned or perceived individual spirit forms with whom magicians can work to achieve Pete's definition of magic itself: "the raising of the whole individual in perfect balance to the power of infinity."

Instead of prayer, or the manipulation of complex natural correspondences, the fundamental motor of empowerment for practitioners consists of meditative exercises to free the subconscious and charge up the will. In pursuit of their goals, they are entitled to draw eclectically on such previous belief and symbolic systems as they please, as long as they remain in control. The prime cautionary injunction is that nothing is true (which presumably means that the injunction itself need not be either).

It remains to be seen what the new century will eventually produce as a successor to Peter Carroll and his system to take forward the theorization of magic. All that a historian can predict is that such a thinker will once more draw on previous traditions and on contemporary cultural developments to produce a distinctive new blend, and that, as it has always done, magic itself will continue to survive and evolve as a human preoccupation.

—Ronald Hutton, author of *The Triumph of the Moon: A History of Modern Pagan Witchcraft* and other works

Introduction

Magic is an intensely practical, personal, experimental art. Two major themes run through this book: that altered states of consciousness are the key to unlocking one's magical abilities; and that these abilities can be developed without any symbolic system except reality itself.

The magical style of thinking is explored with chapters on random belief and the alphabet of desire.

A natural inclination toward the darker side of magic is as good a point as any from which to begin the ultimate quest, and half this book is devoted to the black arts.

Independent of ancient dusty books and mystification, the vital elements of many traditions conspire here to create a living art.

Chaos magic takes its inspiration from the practical techniques of many historical traditions from ancient shamanism to classical pagan magic to more modern traditions such as the Golden Dawn, the Crowleyan Argentum Astrum, and the Zos Kia Cultus of Austin Spare.

This book, written originally as a sourcebook for a secretive group of chaos magicians, is now being released for those who wish to work alone or who wish to establish working chaos magic groups.

The word "magician," like the word "witch," can apply to persons of any gender.

The animism and shamanism of preagricultural societies underlie all magical traditions. They deal with the "forces" and "spirits" associated with natural phenomena, plants, animals, and people.

As humans developed agriculture, settled civilizations, and organized philosophies, religions, and writing, they also developed more complicated ideas about magic that often involved more abstract spirits and deities; and they started to record them. Four written sources of magic seem to have had a particularly lasting impact.

Ancient Egyptian magic influenced the magical traditions of the pagan Greco-Roman (Hellenic) world and the magical ideas in the Abrahamic monotheisms of Judaism, Christianity, and Islam, but perhaps less so than the magical ideas that came from the ancient Persian Empire that practiced

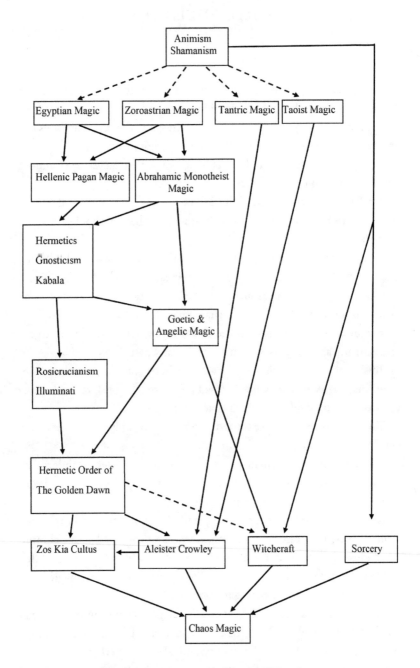

Diagram 1. *The survival of the magical tradition.*

Zoroastrianism. The words "magic" and "magus" both derive from the *magi*, the priest-magicians of ancient Zoroastrianism.

Tantric magic developed alongside the religions of India, and it includes far more than just the use of sex in both its Hindu and Buddhist forms, for it involves all the senses and facilities and subtle energies of the body. Yoga, in all its forms, constitutes a manifestation of Tantra. The Taoism of Chinese civilization developed rich undercurrents of magic. Alchemy, astrology, esoteric martial arts and medicine, feng shui, and sorcery all flourished under the umbrella of Taoism.

Yet the magical traditions of the East, from India and China, began to profoundly influence Western occult thinking only during and after the colonial period.

In the West, the attempt to integrate the magical ideas of Hellenic paganism with those of Judaic monotheism led to the flowering of the Neoplatonic doctrines of Hermetics, Gnosticism, and Kabbala in the first two centuries AD. Much of traditional Western esoterics and magic still uses ideas forged in this period.

All three of the Abrahamic monotheisms developed remarkably similar forms of magic. They all had their own equivalents of angels, demons, and spirits, and they all had similar Neoplatonic ideas about natural magic and natural phenomena. Most of the demons of monotheist magic were drawn from their previous pagan pantheons. Ideas and manuscripts passed back and forth between these cultures over the centuries. During the murkier periods of the Dark and Middle Ages when writing and learning were in short supply, many people fell back on traditional and perhaps instinctive ideas of sorcery derived from animism.

The Renaissance saw a considerable flowering of "learned" magic based on the rediscovery and reinvention of 1st- and 2nd-century magic and esoterics, plus some exclusively monotheis-style magic. Various Illuminati and Rosicrucian initiatives devolve from this period and its aftermath; we shall perhaps never know the real extent of such secretive brotherhoods of magicians or their effect on the Enlightenment.

Meanwhile, the basic idea of animism and shamanism—that natural phenomena, plants, animals, and people have associated forces and spirits that we can interact with—never goes away and persists as a belief in sorcery. Arguably, witchcraft has never existed in an organized form until very recently. Accusations of witchcraft have, however, persisted for millennia against those suspected of using sorcery for evil ends. Contemporary self-professed witches tend to take their ideas from a mixture of animist-inspired sorcery and the "learned" traditions of magic, plus, ironically, the fantasies of their historical accusers about organized witchcraft.

In the late 19th century, a great synthesis of magical ideas came together to create the Magical Order of the Golden Dawn and a huge revival in magic, just as the Industrial Revolution went into overdrive. MacGregor Mathers, and perhaps his wife Moina (the sister of Henri Bergson, the radical philosopher), may well have created nearly all of it. We simply do not know; the GD manuscripts remain unattributed, and biographical data remains sketchy.

While the GD corpus can at first appear as a sprawling attempt at a syncretism of a confusing mass of traditions and symbols, it really consists of a system of practical techniques such as visualization, meditation, incantation, subliminalization, and vibration that can breathe magic into any form of enchantment, divination, invocation, evocation, or illumination from any tradition. It inspired the work of Aleister Crowley and Austin Spare; modern Druidry, Wicca, and neo-paganism; and chaos magic.

Aleister Crowley played a pioneering role in popularizing tantric and Taoist magical techniques in Western magic. His sometime pupil Austin Spare discovered forms of practical magic that dispensed with virtually all supposedly ancient symbols and ideas and used the contents of the personal subconscious instead. This forms the basis of what Spare called the Zos Kia Cultus.

The techniques, rather than the dogma, of all the above traditions have fed into chaos magic to recreate magic as a practical skill for the contemporary world.

LIBER NULL

The Quest

The secrets of magic are universal and of such a practical physical nature as to defy simple explanation. Those beings who realize and practice such secrets are said to have achieved mastership. Masters will, at various points in history, inspire adepts to create magic, mystic, religious, or even secular orders to bring others to mastership. Such orders have at certain times openly called themselves the Illuminati; at other times secrecy has seemed more prudent. The mysteries can be preserved only by constant revelation. In this, chaos magic continues a tradition perhaps ten thousand years old, yet chaos magic has no history. Every culture and generation rewrites the eternal verities of magic in terms of its own symbolism and idiosyncrasies; only the underlying practical techniques of magic really matter.

In the tradition with no past, there is nowhere to conceal the future from the present. It takes its inspiration from sex and death. Apart from being humanity's two greatest obsessions and motivating forces, sex and death represent the positive and negative methods of attaining magical consciousness. Illumination refers to the inspiration, enlightenment, and liberation resulting from success with these methods.

The specific purpose for which the chaos tradition of magic is constituted is to help determine in what form the as yet embryonic fifth aeon will manifest. Its task, although historic, consists of disseminating magical knowledge to individuals. For at no time since the first aeon has humanity stood in such need of these abilities to see its way forward.

There is no formal hierarchy in the chaos tradition of magic. There is a division of activity depending on ability as it develops.

Neophytes strengthen their magical will against the strongest possible adversary—their own minds. They explore the possibilities of changing themselves at will and explore their own occult abilities in dream and magical activity.

Initiates familiarize themselves with all forms of occult attainment and seek to perfect themselves in some particular form of magic. They should also find others capable of aspiring to chaos magic and offer them help.

Adepts seek perfection in all aspects of personal magical power, wisdom, and liberation.

Masters seek to realize the aims of the initiates in chaos magic tradition by whatever forms of action or nonaction they deem appropriate.

Diagram 1 is an exposition of the survival of magical traditions from the first aeon to the fifth. For an extended discussion of the aeonics involved, consult "The Millennium" on page 77.

LIBER MMM

The Neophyte Syllabus
of Chaos Magic

This course is an exercise in the disciplines of magical trance, a form of mind control having similarities to yoga, personal metamorphosis, and the basic techniques of magic. Success with these techniques is a prerequisite for any real progress with the initiate syllabus.

A magical diary is the magician's most essential and powerful tool. It should be large enough to allow a full page for each day. Students should record the time, duration, and degree of success of any practice undertaken. They should make notes about environmental factors conducive (or otherwise) to the work.

Neophytes who wish to study under the supervision and encouragement of experienced chaos magicians willing to take on apprentices are encouraged to do so. Both parties must evolve strategies for finding each other.

Mind Control

TO WORK MAGIC EFFECTIVELY, the ability to concentrate the attention must be built up until the mind can enter a trancelike condition. This is accomplished in a number of stages: absolute motionlessness of the body, regulation of the breathing, stopping of thoughts, and magical trances (concentration on objects, concentration on sound, and concentration on mental images).

Motionlessness

Arrange the body in any comfortable position and try to remain in that position for as long as possible. Try not to blink or move the tongue or fingers or any part of the body at all. Do not let the mind run away on long trains of thought but rather observe oneself passively. What appeared to be a comfortable position may become agonizing with time, but persist! Set aside some time each day for this practice and take advantage of any opportunity of inactivity which may arise.

Record the results in the magical diary. One should not be satisfied with less than five minutes. When fifteen have been achieved, proceed to regulation of the breathing.

Breathing

Stay as motionless as possible and begin to deliberately make the breathing slower and deeper. The aim is to use the entire capacity of the lungs but without any undue muscular effort or strain. The lungs may be held full after inhalation for a few moments, or empty after exhalation for a few moments to lengthen the cycle. The important thing is that the mind should direct its complete attention to the breath cycle. When this can be done for thirty minutes, proceed to "not-thinking."

Not-Thinking

The exercises of motionlessness and breathing may improve health, but they have no other intrinsic value aside from being a preparation for not-thinking, the beginnings of the magical trance condition. While motionless and breathing deeply, begin to withdraw the mind from any thoughts which arise. The attempt to do this inevitably reveals the mind to be a raging tempest of activity. Only the greatest determination can win even a few seconds of mental silence, but even this is quite a triumph. Aim for complete vigilance over the arising of thoughts and try to lengthen the periods of total quiescence.

Like the physical motionlessness, this mental motionlessness should be practiced at set times and also whenever a period of inactivity presents itself. The results should be recorded in your diary.

The Magical Trances

Magic is the science and art of causing change to occur in conformity with will. The will can become magically effective only when the mind is focused and not interfering with the will. The mind must first discipline itself to focus its entire attention on some meaningless phenomenon. If an attempt is made to focus on some form of desire, the effect is short-circuited by lust of result. Egotistical identification, fear of failure, and the reciprocal desire not to achieve desire, arising from our dual nature, destroy the result.

Therefore, when selecting topics for concentration, choose subjects of no spiritual, egotistical, intellectual, emotional, or useful significance—meaningless things.

OBJECT CONCENTRATION

The legend of the evil eye derives from the ability of wizards and sorcerers to give a fixed dead stare. This ability can be practiced on any object—a mark on a wall, something in the distance, a star in the night sky—anything. To hold an object with an absolutely fixed, unwavering gaze for more than a few moments proves extraordinarily difficult, yet it must be maintained for hours at a time. Every attempt by the eye to distort the object, every attempt by the mind to find something else to think of, must be resisted. Eventually it is possible to extract occult secrets from things by this technique, but the ability must be developed by working with meaningless objects.

SOUND CONCENTRATION

The part of the mind in which verbal thoughts arise is brought under magical control by concentration on sounds mentally imagined. Any simple sound of one or more syllables is selected, for example, *Aum* or *Om, Abrahadabra, Yod He Vau He, Aum Mani Padme Hum, Zazas Zazas, Nasatanada Zazas.* The chosen sound is repeated over and over in the mind to block all other thoughts. No matter how inappropriate the choice of sound may seem to have been, you must persist with it. Eventually the sound may seem to repeat itself automatically and may even occur in sleep. These are encouraging signs. Sound concentration is the key to words of power and certain forms of spell casting.

IMAGE CONCENTRATION

The part of the mind in which pictorial thoughts arise is brought under magical control by image concentration. A simple shape, such as a triangle, circle, square, cross, or crescent, is chosen and held in the mind's eye, without distortion, for as long as possible. Only the most determined efforts are likely to make the imagined form persist for any time. At first, the image should be sought with the eyes closed. With practice, it can be projected onto any blank surface. This technique is the basis of casting sigils and creating independent thought forms.

The three methods of attaining magical trance will yield results only if pursued with the most fanatical and morbid determination. These abilities are highly abnormal and usually inaccessible to human consciousness, as they demand such inhuman concentration, but the rewards are great. In the magical diary, record each day's formal work and whatever extra opportunities have been utilized. No page should be left blank.

METAMORPHOSIS

The transmutation of the mind to magical consciousness has often been called the great work. It has a far-reaching purpose leading eventually to the discovery of the true will. Even a slight ability to change oneself is more valuable than any power over the external universe. Metamorphosis is an exercise in willed restructuring of the mind.

All attempts to reorganize the mind involve a duality between conditions as they are and the preferred condition. Thus, it is impossible to cultivate any virtue like spontaneity, joy, pious pride, grace, or omnipotence without involving oneself in more conventionality, sorrow, guilt, sin, and impotence in the process. Religions are founded on the fallacy that one can or ought to have one without the other. High magic recognizes the dualistic condition but does not care whether life is bittersweet or sweet and sour; rather it seeks to achieve any arbitrary perceptual perspective at will.

Any state of mind might arbitrarily be chosen as an objective for trans-mutation, but there is a specific virtue to the ones given. The first is an antidote to the imbalance and possible madness of the magical trance. The second is a specific against obsession with the magical practices in the third section. They are:

1. Laughter/Laughter

2. Non-attachment/Non-disinterest

Attaining these states of mind is accomplished by a process of ongo-ing meditation. One tries to enter into the spirit of the condition when-ever possible and to think about the desired result at other times. By this method, a strong new mental habit can be established.

Consider laughter: it is the highest emotion, for it can contain any of the others from ecstasy to grief. It is its own opposite. Crying is merely an underdeveloped form of it which cleanses the eyes and summons assis-tance to infants. Laughter is the only tenable attitude in a universe which is a joke played upon itself. The trick is to see that joke played out even in the neutral and ghastly events which surround one. It is not for us to question the universe's apparent lack of taste. Seek the emotion of laughter at what delights and amuses, seek it in whatever is neutral or meaningless, seek it even in what is horrific and revolting. Though it may be forced at first, one can learn to smile inwardly at all things.

Non-attachment/non-disinterest best describes the magical condition of acting without lust of result. It is very difficult for humans to decide on something and then to do it purely for its own sake. Yet it is precisely this ability which is required to execute magical acts. Only single-pointed awareness will do. Attachment is to be understood in both the positive and negative sense, for aversion is its other face. Attachment to any attribute of oneself, personality, one's ambitions, one's relationships or sensory experi-ences—or equally, aversion to any of these—will prove limiting.

On the other hand, it is fatal to lose interest in these things for they are one's symbolic system or magical reality. Rather, one is attempting to touch the sensitive parts of one's reality more lightly in order to deny the spoiling hand of grasping desire and boredom. Thereby one may gain enough freedom to act magically.

In addition to these two meditations there is a third, more active, form of metamorphosis, and this involves one's everyday habits. However innocuous they might seem, habits in thought, word, and deed are the anchor of the personality. The magician aims to pull up that anchor and cast himself free on the seas of chaos.

To proceed, select any minor habit at random and delete it from your behavior; at the same time, adopt any new habit at random. The choices should not involve anything of spiritual, egocentric, or emotional significance, nor should you select anything with any possibility of failure. By persisting with such simple beginnings, you become capable of virtually anything

All works of metamorphosis should be committed to the magical diary.

Magic

SUCCESS IN THIS PART OF THE SYLLABUS is dependent on some degree of mastery of the magical trances and metamorphosis. This magical instruction involves three techniques: ritual, sigils, and dreaming. In addition, the magician should become familiar with at least one system of divination: cards, crystal gazing, rune sticks, pendulum, or divining rod. The methods are endless. With all techniques, aim to silence the mind and let inspiration provide some sort of answer. Whatever symbolic system or instruments are used, they act only to provide a receptacle or amplifier for inner abilities. No divinatory system should involve too much randomness. Astrology is not recommended.

Ritual is a combination of the use of talismanic weapons, gesture, visualized sigils, word spells, and magical trance. Before proceeding with sigils or dreaming, it is essential to develop an effective banishing ritual. A well-constructed banishing ritual has the following effects. It prepares the magician more rapidly for magical concentration than any of the trance exercises alone. It enables the magician to resist obsession if problems are encountered with dream experiences or with sigils becoming conscious. It also protects the magician from any hostile occult influences which may assail him.

To develop a banishing ritual, first acquire a magical weapon—a sword, a dagger, a wand, or perhaps a large ring. The instrument should be something which is impressive to the mind and should also represent the aspirations of the magician. The advantages of hand forging one's own instruments, or discovering them in some strange way, cannot be overemphasized. The banishing ritual should contain the following elements as a minimum.

First, the magician describes a surrounding barrier with the magical weapon. The barrier is also strongly visualized. Three-dimensional figures are preferable. See figure 1 on page 12.

Second, the magician focuses attention on a visualized image: for example, the image of the magical weapon, or a personal imaginary third eye, or perhaps a ball of light inside the head. A sound concentration may additionally or alternatively be used.

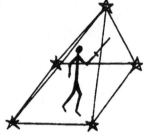

Figure 1. Different forms of three-dimensional barriers that the magician can create using the magical weapon.

Third, the barrier is reinforced with power symbols drawn with the magical weapon. The traditional five-pointed star or pentagram can be used, or the eight-pointed star of chaos, or any other form. Words of power may also be used.

Fourth, the magician aspires to the infinite void by a brief but determined effort to stop thinking.

Sigils

The magician may require something which is unobtainable through the normal channels. It is sometimes possible to bring about the required coincidence by the direct intervention of the will, provided that this does not put too great a strain on the universe. The mere act of wanting is rarely effective, as the will becomes involved in a dialogue with the mind. This dilutes magical ability in many ways. The desire becomes part of the ego complex; the mind becomes anxious of failure. The will not to fulfill desire arises to reduce fear of failure. Soon the original desire is a mass of conflicting ideas. Often the wished-for result arises only when it has been forgotten. This last fact is the key to sigils and most forms of magic spell. Sigils work because they stimulate the will to work subconsciously, bypassing the mind.

There are three parts to the operation of a sigil. The sigil is constructed, the sigil is lost to the mind, the sigil is charged. In constructing a sigil, the aim is to produce a glyph of desire, stylized so as not to immediately suggest the desire. It is not necessary to use complex symbol systems. Figure 2 shows how sigils may be constructed from words, from images, and from sounds. The subject matter of these spells is arbitrary and not recommended.

12

A) *Word method.* I wish to obtain the Necronomicon

INISHTOOBTAINTHENECRONOMICON

(Eliminate repeated letters)

INSHTOBANECRM

Letters rearranged
to give pictorial
sigil

B) *Pictorial method, to restrain adversary*

Finished
sigil

C) *Mantrical spell method*

I want to meet a succubus in dream
IWAH N'AR MEDAR SUKU BUS'X DREEM
IWAH N'MER D'SUK

(Rearrange)

HAWI EMNER KUSAD

Finished
mantra

*Figure 2. Creating a sigil by (A) the word method, (B) the pictorial method,
and (C) the mantrical method.*

To successfully lose the sigil, both the sigil form and the associated desire must be banished from normal waking consciousness. The magician strives against any manifestation of either by a forceful turning of his attention to other matters. Sometimes the sigil may be burnt, buried, or cast into an ocean. It is possible to lose a word spell by constant repetition, as this eventually empties the mind of associated desire.

The sigil is charged at moments when the mind has achieved quiescence through magical trance, or when high emotionality paralyzes its normal functioning. At these times, the sigil is concentrated upon, either as a mental image, or mantra, or as a drawn form. Some of the times when sigils may be charged are as follows: during magical trance; at the moment of orgasm or great elation; at times of great fear, anger, or embarrassment; or at times when intense frustration or disappointment arises. Alternatively, when another strong desire arises, this desire is sacrificed (forgotten) and the sigil is concentrated on instead. After holding the sigil in the mind for as long as possible, it is wise to banish it by evoking laughter.

A record should be kept of all work with sigils but not in such a way as to cause conscious deliberation over the sigilized desire.

Dreaming

THE DREAM STATE PROVIDES a convenient egress into the fields of divination, entities, and exteriorization or "out of the body" experience. All humans dream each night of their lives, but few can regularly recount their experiences even a few minutes after waking. Dream experiences are so incongruous that the brain learns to prevent them interfering with waking consciousness. The magician aims to gain full access to the dream plane and to assume control of it. The attempt to do this invariably involves the magician in a deadly and bizarre battle with his own psychic censor, which will use almost any tactics to deny him these experiences.

The only method of gaining full access to the dream plane is to keep a book and writing instrument next to the place of sleeping at all times. In this, record the details of all dreams as soon as possible after waking.

To assume conscious control over the dream state, it is necessary to select a topic for dreaming. The magician should start with simple experiences, such as the desire to see a particular object (real or imaginary). and master this before attempting divination or exteriorization. The dream is set up by strongly visualizing the desired topic in an otherwise silenced mind immediately before sleep. For more complex experiences, the method of sigils may be employed.

A record of dreams is best kept separate from the magical record, as it tends to become voluminous. However, any significant success should be transferred into the magical diary.

Though one may get to fear the sight of it, a properly kept magical record is the surest guarantor of success in the work of *Liber MMM*: it is both a work of reference with which to evaluate progress and, most significantly, a goad to further effort.

LIBER LUX

Initiate Syllabus 1

T he magic art may be subdivided in many ways: by the ethical tone of the intent, by the moralistic qualities of the effects, into high and low, and so on. The division favored here is more temperamental. White magic leans more toward the acquisition of wisdom and a general feeling of faith in the universe. The black form is concerned more with the acquisition of power and is reflective of a basic faith in oneself. The end results are likely to be not dissimilar, for the paths crisscross over each other in strange ways.

Initiates are at liberty to work with material from either or both. The so-called middle way, or path of knowledge, consisting of the acquisition of secondhand ideas, is an excuse to do neither and leads nowhere.

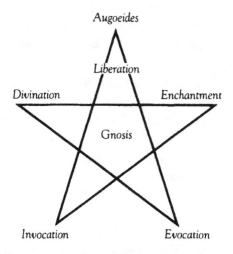

Figure 3. The schema of Liber Lux.

Being the initiate syllabus in white magic, the subject is divided in accordance with the schema shown in figure 3, and we discuss the theoretical aspects of chaos, Kia, duality, aether, and mind.

Duality describes humanity's usual condition. Happiness exists only because of misery, pain because of comfort, good because of evil, yang because of yin, black because of white, birth because of death, and existence because of nonexistence. All phenomena must be paired, as the senses are only equipped to perceive differences. The thinking mind has the property of splitting everything it encounters into two, as it is a dualistic thing itself. Yet there is a part of man which is of a singular nature, although the mind is unable to perceive it as such. Man considers himself a center of will and a center of perception. Will and perception are not separate but only appear so to the mind. The unity which appears to the mind to exert the twin functions of will and perception is called *Kia* by magicians. Sometimes it is called the spirit, or soul, or life force, instead.

Kia cannot be experienced directly because it is the basis of consciousness (or experience), and it has no fixed qualities which the mind can latch on to. Kia is the consciousness; it is the elusive "I" which confers self-awareness but does not seem to consist of anything itself. Kia can sometimes be felt as ecstasy or inspiration, but it is deeply buried in the dualistic mind. It is mostly trapped in the aimless wanderings of thought and in identification with experience and in that cluster of opinions about ourselves called ego. Magic is concerned with giving the Kia more freedom and flexibility and with providing means by which it can manifest its occult power. Kia is capable of occult power because it is a fragment of the great life force of the universe.

Consider the world of apparent dualisms we inhabit. The mind views a picture of this world in which everything is double. A thing is said to exist and exert certain properties. "Being" and "doing." This calls for the concepts of cause and effect, or causality. Every phenomenon is seen to be caused by some previous thing. However, this description cannot explain how everything exists in the first place or even how one thing finally causes another. Obviously, things have originated and do continue to make one another happen. The "thing" responsible for the origin and continued action of events is called *chaos* by magicians. It could as well be called God or Tao, but the name "Chaos" is virtually meaningless, and free from the childish, anthropomorphic ideas of religion.

Chaos is also the force which adds increasing complexity to the universe by spawning structures which were not inherent in its component parts. It is the force which has caused life to evolve itself out of dust and is currently most concentratedly manifest in the human life force, or Kia, where it is the source of consciousness.

Kia is but a small fragment of the great life force of the universe, which contains the twin impulses to immerse itself in duality and to escape from duality. It will continuously reincarnate until the first impulse is exhausted. The second impulse is the root of the mystic quest, the union of the liberated spirit with the great spirit. To the extent that the Kia can become one with chaos, it can extend its will and perception into the universe to accomplish magic.

Between chaos and ordinary matter, and between Kia and the mind, there exists a realm of half-formed substance called *aether*. It is dualistic matter but of a very tenuous, probabilistic nature. It consists of all the possibilities which chaos throws out which have not yet become solid realities. It is the "medium" by which the "nonexistent" chaos translates itself into "real" effects. It forms a sort of backdrop out of which real events and real thoughts materialize. Because aetheric events are only partially evolved into dualistic existence, they may not have a precise location in space or time. They may not have a precise mass or energy either, and so do not necessarily affect the physical plane.

It is from the bizarre and indeterminate nature of the aetheric plane that chaos gets its name, for chaos cannot be known directly.

From the aetheric realm of nascent possibility, only what we call sensible, causal, probable, or normal events usually come into existence. Yet as centers of Kia or chaos ourselves, we can sometimes call very unlikely coincidences or unexpected events into existence by manipulating the aether. Such is magic. Even the physical sciences have begun to blunder into the aetheric with their discoveries of quantum indeterminacy and virtual processes in subatomic matter.

It is the aether, which surrounds the central core of the life force, with which the magician is concerned. Its normal function is as a Kia-thought intermediary, yet its properties are so infinitely mutable that almost anything can be accomplished with it. Thought gives it shape, and Kia gives it power.

Thus are will and perception extended into areas of time and space beyond the physical limitations of the material body.

It is the very mutability of the aetheric which has given rise to such a bewildering variety of magical activity and supportive thought forms all over the universe. The differences, however, are only superficial. When stripped of local symbolism and terminology, all systems show a remarkable uniformity of method. This is because all systems ultimately derive from the traditions of animism and shamanism.

It is toward an elucidation of this tradition that the following chapters are devoted.

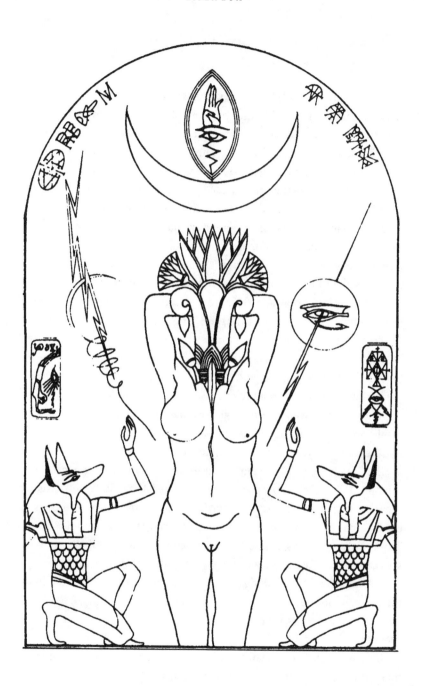

Gnosis

ALTERED STATES OF CONSCIOUSNESS are the key to magical powers. The particular state of mind required has a name in every tradition: No-mind. Stopping the internal dialogue, passing through the eye of the needle, *ain* or nothing, *samadhi,* or one-pointedness. In this book, it will be known as *gnosis.* It is an extension of the magical trance by other means.

Methods of achieving gnosis can be divided into two types. In the inhibitory mode, the mind is progressively silenced until only a single object of concentration remains. In the excitatory mode, the mind is raised to a very high pitch of excitement while concentration on the objective is maintained. Strong stimulation eventually elicits a reflex inhibition and paralyzes all but the most central function—the object of concentration. Thus, strong inhibition and strong excitation end up creating the same effect—the one-pointed consciousness, or gnosis.

Neurophysiology has finally stumbled on what magicians have known by experience for millennia. As a great master once observed: "There are two methods of becoming god, the upright or the averse." Let the mind become as a flame or a pool of still water. It is during these moments of single-pointed concentration, or gnosis, that beliefs can be implanted for magic, and the life force induced to manifest. Table 1 on page 23 shows a number of methods that can be used to attain it.

The "death posture" is a feint at death to achieve an utter negation of thought. It can take many forms, ranging from the simple not-thinking exercise up to complex rituals. A very fast and simple method consists of blocking the ears, nose, and mouth, and covering the eyes with the hands. The breath and thoughts are forcefully jammed back until near unconsciousness involuntarily breaks the posture. Alternatively, one may arrange oneself before a mirror at a distance of about two feet and stare fixedly at the image of one's eyes in the mirror with an unblinking, corpse-like gaze. The effort required to keep an absolutely unwavering image will of itself silence the mind after a while.

Sexual excitation can be obtained by any preferred method. In all cases there has to be a transference from the lust required to ignite the sexuality to the matter of the magical working at hand.

Table 1. The Physiological Gnosis

Inhibitory Mode	Excitatory Mode
Death posture	Sexual excitation
Magical trance Concentration	Emotional arousal, e.g., fear, anger, and horror
Sleeplessness Fasting Exhaustion	Pain, torture Flagellation Dancing, drumming Chanting
Gazing	Right way of walking
Hypnotic or trance-inducing drugs	Excitatory or disinhibitory drugs, mild hallucinogens, forced over-breathing
Sensory deprivation	Sensory overload

The nature of a sexual working lends itself readily to the creation of independent orders of being—evocation. Also, in works of invocation where the magician seeks union with some principle (or being), the process can be mirrored on the physical plane; one's partner is visualized as an incarnation of the desired idea or god. Prolonged sexual excitement through karezza, inhibition of orgasm, or repeated orgasmic collapse can lead to trance states useful for divination. It may be necessary to regain one's original sexuality from the mass of fantasy and association into which it mostly sinks. This is achieved by judicious use of abstention and by arousing lust without any form of mental prop or fantasy. This exercise is also therapeutic. Be ye ever virgin unto Kia.

The concentrations leading to magical trance are discussed in *Liber MMM*. Emotional arousal is the obverse form of this method. Emotive arousal of any sort can theoretically be used, even love or grief in extreme circumstances, but in practice only anger, fear, and horror can easily be generated in sufficient strengths to achieve the requisite effect. The well-known ability of fear and anger to paralyze the mind indicates their effectiveness, yet the magician must never lose sight of the objectives of his working. Nothing is to be gained and much may be lost by reducing oneself to gibbering idiocy or catatonia.

Sleeplessness, fasting, and exhaustion are old monastic favorites. There should be a constant turning of the mind toward the object of the exercise during these practices. Pain, torture, and flagellation have been used by witches, monks, and fakirs to achieve results. Surrender to pain brings eventual ecstasy and the necessary one-pointedness. However, if the organism's resistance to pain is high, needless damage to the body may result before the threshold is crossed.

Dancing, drumming, and chanting require careful arranging and preparation to bring the participants to a climax. Lyrical exaltation through emotive poetry, incantation, song, prayer, or supplication can also be added. The whole is best controlled by some form of ritual. Over-breathing is sometimes used to supplement the effects of dancing or leaping.

The right way of walking is not a technique for achieving immediate results but a meditation which helps the mind to stop thinking. One walks for long stretches without looking at anything directly but, by slightly crossing and unfocusing the eyes, maintaining a peripheral view of everything. It should be possible to remain cognizant of everything within a 180-degree arc from side to side and from the tips of the toes to the sky. The fingers should be curled or clasped in unusual positions to draw attention to the arms. The mind should eventually become totally absorbed in its environment and thinking will cease.

Gazing is the inhibitory variant of the above technique. The entire attention is directed to the sight of some object in the environment while the body is kept motionless. Any natural phenomenon—plants, rocks, sky, water, or fire—may be used.

There is no magic drug which will, by itself, have the required effect. Rather drugs can be used in small doses to heighten the effect of excitation caused by the methods already discussed. In all cases a large dose leads to depression, confusion, and a general loss of control. Inhibitory drugs must be considered with even more caution because of their inherent danger. They often simply sever the life force and body altogether.

Sensory overload is achieved when a battery of techniques are used together. For example, in certain tantric rites, the candidate is first beaten by his guru, hashish is forced down him, and he is taken at midnight to a dark cemetery for sacred sexual intercourse. Thus, he achieves union with his god.

Sensory deprivation is the essence of the monastic cell, the mountain cave, the walled-up hermit, and rites of death, burial, and resurrection. Much the same effect can be achieved with hoods, blindfolds, earplugs, repetitive sounds, and restricted movements. It is far more effective to completely obliterate all sensory inputs for a short period than to simply reduce them over a longer one.

Certain forms of gnosis lend themselves more readily to some forms of magic than others. The initiate is encouraged to use his own ingenium in adapting the methods of exaltation to his own purposes.

Note, however, that inhibitory and excitatory techniques can be employed sequentially, but not simultaneously, in the same operation.

Evocation

EVOCATION IS THE ART OF DEALING with magical beings or entities by various acts which create or contact them and allow one to conjure and command them with pacts and exorcism. These beings have a legion of names drawn from the demonology of many cultures: elementals, familiars, incubi, succubae, bud-wills, demons, automata, atavisms, wraiths, spirits, and so on. Entities may be bound to talismans, places, animals, objects, persons, or incense smoke or be mobile in the aether. It is not the case that such entities are limited to obsessions and complexes in the human mind. Although such beings customarily have their origin in the mind, they may be budded-off and attached to objects and places in the form of ghosts, spirits, or "vibrations," or may exert action at a distance in the form of fetishes, familiars, or poltergeists. These beings consist of a portion of Kia, or the life force, attached to some aetheric matter, the whole of which may or may not be attached to ordinary matter.

Evocation may be further defined as the summoning or creation of such partial beings to accomplish some purpose. They may be used to cause change in oneself, change in others, or change in the universe. The advantages of using a semi-independent being rather than trying to effect a transformation directly by will are several: the entity will continue to fulfill its function independently of the magician until its life force dissipates. Being semi-sentient, it can adapt itself to a task in a way that a nonconscious simple spell cannot. During moments of the possession by certain entities, the magician may be the recipient of inspirations, abilities, and knowledge not normally accessible to him.

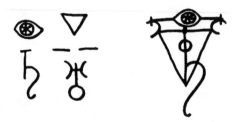

Figure 4. Creating an elemental by combining appropriate symbols to form a sigil.

26

Entities may be drawn from three sources—those which are discovered clairvoyantly, those whose characteristics are given in grimoires of spirits and demons, and those which the magician may wish to create.

In all cases, establishing a relationship with the spirit follows a similar process of evocation. Firstly, the attributes of the entity, its type, scope, name, appearance, and characteristics, must be placed in the mind or made known to the mind. Automatic drawing or writing, where a stylus is allowed to move under inspiration across a surface, may help to uncover the nature of a clairvoyantly discovered being. In the case of a created being, the following procedure is used: the magician assembles the ingredients of a composite sigil of the being's desired attributes. For example, to create an elemental to assist with divination, the appropriate symbols might be chosen and made into a sigil such as the one shown in figure 4.

A name and an image and, if desired, a characteristic number can also be selected for the elemental.

Secondly, the will and perception are focused as intently as possible (by some gnostic method) on the elemental's sigils or characteristics so that these take on a portion of the magician's life force and begin autonomous existence. In the case of pre-existing beings, this operation serves to bind the entity to the magician's will.

This is customarily followed by some form of self-banishing, or even exorcism, to restore the magician's consciousness to normal before going forth.

An entity of a low order with little more than a singular task to perform can be left to fulfill its destiny with no further interference from its master. If at any time it is necessary to terminate it, its sigil or material basis should be destroyed and its mental image destroyed or reabsorbed by visualization. For more powerful and independent beings, the conjuration and exorcism must be in proportion to the power of the ritual which originally evoked them. To control such beings, the magicians may have to re-enter the gnostic state to the same depth as before in order to draw their power.

Any of the techniques of the gnosis can in theory be used in evocation. An analysis of some of the more common methods follows.

Theurgic ritual depends solely on visualization and concentration on complex ceremonies to achieve focus. However, the effect of increasing the complexity is often to create more distraction rather than draw attention to the matter at hand. Will becomes multiple, and the result is often disappointing. Conjuration by prayer, supplication, or command is rarely effective unless the appeal is desperate or prolonged until exhaustion ensues. This type of ritual can be improved by the use of poetic exaltation, chanting, ecstatic dancing, and drumming.

The Goetic tradition of the grimoires uses an additional technique. Terror. The grimoires were compiled by Catholic priests, and much of what they wrote was deliberate abomination in their own terms. Transport the whole rite to a graveyard or crypt at midnight, and one has compounded a powerful mechanism for concentrating the Kia by paralyzing the peripheral functions of the mind by fear. If the magicians can maintain control under these conditions, their will is singular and mighty.

The Ophidian tradition uses sexual orgasm to focus the will and perception. It is interesting to note that poltergeist activity invariably centers on the sexually disturbed, usually children at puberty or, more rarely, women at menopause. During these periods of acute tension, intense excitation can channel the mind and allow the life force to manifest frustration outside the body by hurling objects around.

To perform evocation by the Ophidian method, the attributes of an entity in sigilized form are concentrated on at orgasm and may be afterward anointed with the sex fluids. The process is rather like the deliberate creation of an obsession. If enough power can be put into it, it will be capable of independent existence. Incubi and succubae are pre-existing entities created by other peoples' pathological sexuality. Incubi traditionally seek sexual intercourse with living females and succubae with males, often during sleep. However, both forms are almost invariably male, though succubae may make some slight attempt to disguise themselves as females. Unfortunately, they are both predatory and stupid, with little power or motivation for anything but sex.

Sacrifice has been used in the past to create fear or terror, or to invoke the gnosis of pain in support of Goetic-type evocations. However, this method easily exhausts itself, and the sorcerer may end up wading in oceans of blood, much as the Aztecs did, for very little result. Blood sacrifice is most effective and most easily controlled by the use of one's own blood, which is customarily allowed to fall upon the sigil or talisman of the demon. However, the power to control blood sacrifice usually brings with it the wisdom to avoid it in favor of other methods.

Conjuration to visible appearances to prove to oneself, or others, the objective reality of spirits is an ill-considered act. The conditions necessary for its appearance will always allow the retention of the belief that such things are the result of hypnosis, hallucination, or delusion. Indeed, they are a hallucination, for such things do not normally have a physical appearance and have to be persuaded to assume one. Fasting, sleep, and sensory deprivation combined with drugs and clouds of incense smoke will usually provide a demon with sufficiently sensitive and malleable media in which to manifest an image if commanded to do so.

The medieval idea of a pact is an overdramatization, but it contains a germ of truth. All one's thoughts, obsessions, and demons must be reabsorbed before Kia can become one with chaos. However useful such things may be to him in the short term, the sorcerer must eventually recant.

Invocation

THE ULTIMATE INVOCATION, that of Kia, cannot be performed. The paradox is that as Kia has no dualized qualities, there are no attributes by which to invoke it. To give it one quality is merely to deny it another. As an observant dualistic being once said:

I am that I am not.

Nevertheless, magicians may need to make some rearrangements or additions to what they "are" or do. Metamorphosis may be pursued by seeking that which one is not and transcending both in mutual annihilation. Alternatively, the process of invocation may be seen as adding to the magician's psyche any elements which are missing. It is true that the mind must be finally surrendered as one enters fully into chaos, but a complete and balanced psychocosm is more easily surrendered.

The magical process of shuffling beliefs and desires attendant upon the process of invocation also demonstrates that one's dominant obsessions or personality are quite arbitrary, and hence more easily banished.

There are many maps of the mind (psychocosms), most of which are inconsistent, contradictory, and based on highly fanciful theories. Many use the symbology of god forms, for all mythology embodies a psychology. A complete mythic pantheon resumes all of man's mental characteristics. Magicians will often use a pagan pantheon of gods as the basis for invoking some particular insight or ability, as these myths provide the most explicit and developed formulation of the particular idea extant. However, it is possible to use almost anything from the archetypes of the collective unconscious to the elemental qualities of alchemy.

If the magician taps a deep enough level of power, these forms may manifest with sufficient force to convince the mind of the objective existence of the god. Yet the aim of invocation is temporary possession by the god, communication from the god, and manifestation of the god's magical powers, rather than the formation of religious cults.

The actual method of invocation may be described as a total immersion in the qualities pertaining to the desired form. One invokes in every conceivable way. Magicians first program themselves into identity with the god by arranging all their experiences to coincide with its nature. In the most elaborate form of ritual, magicians may surround themselves with the sounds, smells, colors, instruments, memories, numbers, symbols, music, and poetry suggestive of the god or quality. Secondly, they unite their life force to the god image with which they have united their minds. This is accomplished using techniques from the gnosis. Figure 5 shows some examples of maps of the mind. Following are some suggestions for practical ritual invocation.

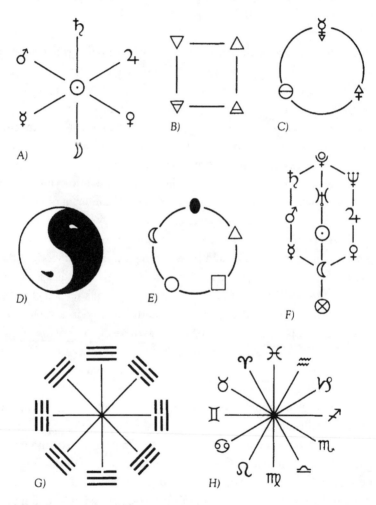

Figure 5. An assortment of psychocosms or mental maps. Magicians may wish to invoke some of the qualities represented by the symbols in each. Here we see (A) the seven classical planetary forms; (B) the four classical elements; (C) the three alchemical elements; (D) the Taoist yin-yang; (E) the five Vedic tattwas; (F) the eleven Kabbalistic sephiroth; (G) the eight Taoist trigrams; and (H) the twelve astrological qualities.

Example Invocations

To evoke the war god Mars, the initiate stands in a pentagonal chamber lit by five red lamps, robed in crimson and the skin of a great bear or wolf, girded about with weapons of steel, and wearing an iron crown or helmet. The magician prepares by fasting, by rigors, by scourging, and by stimulants, and by constantly turning the mind to things of Mars during the preparations.

Cast sulphur, oak, and acris resins into the thurible and anoint the body with tiger balm. Beat a martial air upon a drum to open the temple, or else fire a loud weapon into the air. Banish all foreign influences from the mind by whatever means (a pentagram ritual being preferred).

Drawing blood from the right shoulder with a dagger, trace the sigil of Mars on the breast and the Eye of Horus on the brow. With a sharp sword, draw the symbols of Mars all about in the mind's eye in lines of crimson fire and visualize union with the form of the god Horus.

Then begin the war dance while an assistant continues to beat the rhythm, apply the scourge, or discharge firearms. Martial music may be played by some machine, as the magician dances wildly to the war god, chanting:

Io Horus Horus!
Horus to me come!
GEBURAZARPE!
Thou art me, Horus!
I am thee, Horus!

Note that any of these props can be dispersed by anyone whose Kia flows steadily into the willed artifacts of imagination. There is no limit to the inconceivable experiences into which the intrepid psychonaut may wish to plunge. Here are some ideas for constructing a latter-day black mass as a blasphemy against the gods of logic and rationality. The great mad goddess Chaos, a lower aspect of the ultimate ground of existence in anthropomorphic form, can be invoked for her ecstasy and inspiration.

Drumming, leaping, and whirling in freeform movement are accompanied by idiotic incantations. Forced deep breathing is used to provoke hysterical laughter. Mild hallucinogens and disinhibitory agents (such as alcohol) are taken together with sporadic gasps of nitrous oxide gas. Dice are thrown to determine what unusual behavior and sexual irregularities will take place. Discordant music is played and flashing lights splash onto billowing clouds of incense smoke. A whole maelstrom of ingredients is used to overcome the senses. On the altar, a great work of philosophy, preferably by Russell, lies open, its pages fiercely burning.

Saturn, the god of death, might be invoked in the following manner. First prepare by fasting, sleeplessness, and exhaustion. Retire to a chamber, which is in near total darkness, being illuminated only by three sticks of a resinous, cloying, musty incense. Weigh the body down by wrapping sheets of lead around the limbs, trunk, and head. Otherwise, be cold and naked. To a slow, monotonous drumbeat, conduct a mock burial of oneself. With extreme caution, take small quantities of atropine-like solanum alkaloids. Then meditate on oneself in the aspect of a corpse or skeleton arising slowly from the tomb in a tattered winding sheet and assuming the scythe of office.

The magician may invoke Jupiter for the charisma of leadership and power over others. Such charisma depends upon various tricks but fundamentally upon tricks that the users of such charisma play on themselves. By crowning oneself, one steps toward royalty among men; careful stage management and a judicious escalation of requests for compliance can do the rest. The pretense of self-confidence generates actual self-confidence and with it an aura of confidence that attracts the confidence and compliance of others.

To invoke the Jupiterian powers, magicians arraign themselves with symbols of appropriate power, perhaps scepters and crowns and orbs, or ceremonial weapons, grand headdresses, or symbolic wealth such as enormous checks written out to oneself. To the sound of bombastic and triumphant music, they may march proudly around or adopt dominating postures while resonantly proclaiming their virtues, ambitions, and entitlements to succeed, all the while visualizing their own empowerment and dominion in the world, and the sigil of Jupiter in glowing golds and blues.

Yet after the crescendo, magicians should draw about themselves a veil of false humility before they go forth. Arrogance works best when retained internally rather than openly displayed.

In works of invocation, nothing succeeds like excess.

Liberation

IN CREATING LIFE OUT OF the primeval slime, chaos has always sought to increase its possibilities of expression and to diversify its manifestations. During the evolution of life there have been many stagnant periods and some reversals. But overall, the inherent superiority of the most flexible, adaptable, inclusive, and complex creatures, cultures, men, and ideas always wins out. To seek these qualities is to achieve more liberation than any bizarre feat of renunciation or reorganization of political power is likely to create.

It is a mistake to consider any belief more liberated than another. It is the possibility of change which is important. Every new form of liberation is destined to eventually become another form of enslavement for most of its adherents. There is no freedom from duality on this plane of existence, but one may at least aspire to choose duality.

Liberating behavior is that which increases one's possibilities for future action. Limiting behavior is that which tends to narrow one's options. The secret of freedom is not to be drawn into situations where one's number of alternatives becomes limited or even unitary.

This is an abominably difficult path to tread. It means stepping outside one's own culture, society, relationships, family personality, beliefs, prejudices, opinions, and ideas. It is just these comforting chains which seem to give definition, meaning, character, and a sense of belonging to most people. Yet, in casting off one set of chains, one cannot avoid adopting another set unless one wishes to live in a very reduced and impoverished style—itself a limitation.

The solution is to become omnivorous. Someone who can think, believe, or do any of a half dozen different things is freer and more liberated than someone confined to only one activity.

For this reason, Sufi mystics were required to master a handful of secular trades in addition to their occult studies.

Chief among the techniques of liberation are those which weaken the hold of society, convention, and habit over the initiate, and those which lead to a more expansive outlook.

They are sacrilege, heresy, iconoclasm, bioaestheticism, and anathemism.

Sacrilege: Destroying the Sacred

Energy is liberated when an individual breaks through rules of conditioning with some glorious act of disobedience or blasphemy. This energy strengthens the spirit and gives courage for further acts of insurrection. Put a brick through your television; explore sexualities which are unusual to you. Do something you normally feel to be utterly revolting. You are free to do anything, no matter how extreme, so long as it will not restrict your own or somebody else's future freedom of action.

Heresy: Alternative Definitions

By seeking out ideas which seem bizarre, crazy, extreme, arbitrary, contradictory, and nonsensical, you will find that the ideas you previously clung to as reasonable, sensible, and humanitarian are actually just as bizarre, crazy, and so on. Whatever is suppressed, restricted, ridiculed, or despised almost always contains a telling counterpoint to mainstream ideas. In argument, always disagree, especially if your opponent begins to voice your own opinions.

Iconoclasm: Breaking Images

Immense gulfs exist in human affairs between theory and practice, means and ends. Contrast pornography and romance, cordon bleu gluttony and skeletal famine, dignity and masturbation. Consider violence as entertainment. Mass slaughter for idealism's sake. Look at what goes on in the name of religion and the consumer society. Relish the cacophony of neurosis, fantasy, and psychosis which guides material sensationalist culture to an uncertain end. Picking through society's dirty underwear, we discover its real habits. You can extend this list indefinitely and indeed you should. For human folly is without limit, though society does much to disguise its darker side. Cynicism, sadness, or laughter is the magician's privilege.

Bioaestheticism: The Body

There is a thing more trustworthy than all the sages and which contains more wisdom than a great library. Your own body. It asks only for food, warmth, sex, and transcendence. Transcendence, the urge to become one with something greater, is variously satisfied in love, humanitarian works, or in the artistic, scientific, or magical quests of truth. To satisfy these simple needs is liberation indeed. Power, authority, excessive wealth, and greed for sensory experience are aberrations of these things.

Anathemism: Self-destruction

Sidestepping conventionality still leaves you with a mass of prejudices, idiosyncrasies, identifications, and preferences which give comfort and definition to the personality or ego. An idea cannot be said to be completely understood until you understand the conditions under which it is not true. Similarly, you cannot be said to possess a personality until you are able to manipulate or discard it at will.

Anathemism is a technique practiced directly upon yourself. Eat all loathsome things until they no longer revolt you. Seek union with all that you normally reject. Scheme against your most sacred principles in thought, word, and deed. You will eventually have to witness the loss or putrefaction of every loved thing. Therefore, reflect upon the transitory and contingent nature of all things. Examine everything you believe, every preference, and every opinion, and cut it down.

The personality, a mask of convenience, becomes stuck to the face. Eye becomes clouded by "I." The human spirit becomes a trivial mess of petty identifications. The most cherished principles are the greatest lies. "I think therefore I am." But what is "I"? The more you think, the more the I closes. Thinking, "I am asleep"; my I is blinded. The intellect is a sword, and its use is to prevent identification with any particular phenomenon encountered. The most powerful minds cling to the fewest fixed principles. The only clear view is from atop the mountain of your dead selves.

Augoeides

THE MAGICIAN'S MOST IMPORTANT invocation is that of the personal genius, daemon, true will, or Augoeides. This operation is traditionally known as attaining the knowledge and conversation of the holy guardian angel. It is sometimes known as the magnum opus or great work.

The Augoeides may be defined as the most perfect vehicle of Kia on the plane of duality. As the avatar of Kia on earth, the Augoeides represents the true will, the raison d'être of the magician, the magician's purpose in existing. The discovery of one's true will or real nature may be difficult and fraught with danger, since a false identification leads to obsession and madness.

The operation of obtaining the knowledge and conversation is usually a lengthy one. The magician is attempting a progressive metamorphosis, a complete overhaul of personal existence. Yet the magician has to seek the blueprint for a reborn self—while living. Life is less the meaningless accident it seems. Kia has incarnated in these particular conditions of duality for some purpose. The inertia of previous existences propels Kia into new forms of manifestation. Each incarnation represents a task, or a puzzle to be solved, on the way to some greater form of completion.

The key to this puzzle is in the phenomena of the plane of duality in which we find ourselves. We are, as it were, trapped in a labyrinth or maze. The only thing to do is move about and keep a close watch on the way the walls turn. In a completely chaotic universe such as this one, there are no accidents. Everything is significant. Move a single grain of sand on a distant shore and the entire future history of the world will eventually be changed.

Persons doing their true will are assisted by the momentum of the universe and seem possessed of amazing good luck. In beginning the great work of obtaining the knowledge and conversation, magicians vow "to interpret every manifestation of existence as a direct message from the infinite chaos to themselves personally." To do this is to enter the magical worldview in its totality.

The idea that things happen to one that may or may not be related to the way one acts is an illusion created by our shallow awareness.

Keeping a close eye on the walls of the labyrinth, the conditions of his existence, the magician may then begin the invocation. The genius is not something added to oneself. Rather it is a stripping away of excess to reveal the god within.

Directly on awakening, preferably at dawn, the initiate goes to the place of invocation. Figuring as one goes that being born anew each day brings with it the chance of greater rebirth, first one banishes the temple of the mind by ritual or by some magical trance. Then one unveils some token or symbol or sigil which represents the holy guardian angel. This symbol will likely have to change during the great work as the inspiration begins to move. Next one invokes an image of the angel into the mind's eye. It may be considered as a luminous duplicate of one's own form standing in front of or behind oneself, or simply as a ball of brilliant light above one's head. Then one formulates aspirations in any chosen manner, by humble prayer or exalting proclamation as need be.

The best form of this invocation is spoken spontaneously from the heart and, if halting at first, will prove itself in time. The aim is to establish a set of ideas and images which correspond to the nature of one's genius and at the same time receive inspiration from that source. As the magician begins to manifest truer will, the Augoeides will reveal images, names, and spiritual principles by which it can be drawn into greater manifestation. Having communicated with the invoked form, the magician should draw it into themself and go forth to live in the way they have willed.

The ritual may be concluded with an aspiration to the wisdom of silence by a brief concentration on the sigil of the Augoeides, but never by banishing. Periodically, more elaborate forms of ritual, using more powerful forms of gnosis, may be employed.

At the end of the day, there should be an accounting and fresh resolution made. Though every day be a catalog of failure, there should be no sense of sin or guilt. Magic is the raising of the whole individual in perfect balance to the power of infinity, and such feelings are symptomatic of imbalance.

If any unnecessary or imbalanced scraps of ego become identified with the genius by mistake, then disaster awaits. The life force flows directly into these complexes and bloats them into grotesque monsters, one of which is the demon Choronzon. Some magicians attempting to go too fast with this invocation have failed to banish this demon, and have gone spectacularly insane as a result.

Divination

SPACE, TIME, MASS, AND ENERGY originate from chaos, have their being in chaos, and through the agency of the aether are moved by chaos into the multiple forms of existence.

Some of the various densities of the aether have only a partial or probabilistic differentiation into existence and are somewhat indeterminate in space and time. In the same way that mass exists as a curvature in space-time, extending with a gradually diminishing force to infinity that we recognize as gravity, so do all events, particularly events involving the human mind, send ripples through all creation.

Various methods of intercepting and interpreting these ripples constitute the mantic art or divination. These ripples through space and time can only be received if they strike a note of resonance in the receiver and are not drowned out by noise or suppressed by the psychic censor. Some forms of resonance exist naturally, as between a mother and child, or between lovers. Otherwise, they have to be established by concentrating on the object of divination.

The general level of mental noise can be suppressed by silencing the mind by some gnostic method. This also assists with the concentration. The inhibitory mode of the gnosis is most frequently used. Sleeplessness, fasting, and exhaustion may cause prescience through visions, but as with drugs, there is always the difficulty of maintaining concentration. Any form of magical trance can be adapted for divination by first directing an intense concentration toward the desired matter of divination (or some sigilized form of it) and then allowing impression to arise into the vacuous state of consciousness.

Many of the excitatory techniques can be used, but some with difficulty. Augury may be made by sacrifice, and men have tortured themselves for knowledge, but sex is the easiest. Erotocomatose lucidity (or sex-trance) describes a condition brought about by continually stimulating and exhausting the sexuality by any possible means until the mind enters the borderland state between consciousness and unconsciousness.

So far, only direct prescience, the ideal of divination, has been discussed. This is not always possible, and recourse must often be had to the use of symbolic intermediaries. These can augment the practice of divination greatly or ruin it utterly.

Assuming that the magical perception can forge some sort of tenuous connection with the answer to a question, symbols are shuffled, drawn, or selected in some manner to carry the answer into the conscious mind. Then a further effort must be made in the interpretation to get that magical perception to come into complete manifestation. Symbols are easy to come by; any system can be used—the difficulty lies in forging the magic link. In obtaining the symbolic result, the magician tries to let the magic slip through below the level of conscious control, but must not let the process become merely random. For example, in cartomancy or tarot divination, one should look through the pack first and then shuffle but lightly, or the result will be completely random, and the chances of the spread being able to stimulate the magical perception will be reduced.

Once the symbol has been obtained, it should be used to help the magical perception crystallize more fully. It should become a basis for lateral thinking (or intuitive guesswork) rather than a final answer to be mechanically interpreted.

Astrology is not a valid form of magical divination because it assumes a causal relationship between events which are linked only very weakly if at all. If the relationship were strong, then astrology would be an ordinary secular science. As the relationship is very weak, astrology owes whatever success it has to the natural prescience of its practitioners and obscures its failures with imprecision, evasiveness, and ambiguity.

The best methods of obtaining symbolic intermediate results are those which are just below the threshold of deliberateness but above the threshold of pure randomness. Shamanistic-type methods involving the casting of bones, stones, or sticks marked with runes are simplest and best. As methods involving the fall of coins or dice, the separation of yarrow stalks and their rules for interpretation became progressively more complex the more remote the prescient ability became. Highly complex mathematical systems represent decadence of the art.

Of all the forces which obstruct divination, none has more power over the civilized consciousness than what is called the psychic censor. This is the same factor which denies us access to most of our dream experiences and prevents us from being overwhelmed by the millions of sensory impressions which bombard our body ceaselessly. Although we could not function without it, it is useful to be able to turn parts of it off at times. Hallucinogenic drugs knock it out unselectively and are not much use.

The magician must begin to notice all coincidences which surround him, instead of dismissing them. Often one notices that just before somebody said something, or an event occurred, one knew it would happen. This can happen several times a day, but we somehow, almost unbelievably, manage to dismiss it each time and not connect the occurrences together. If a definite effort is made to consciously note these occurrences, as well as to record them in the magical diary, they start to become much more numerous. So many coincidences occur that it is ridiculous to use the word coincidence at all. One is becoming prescient.

Enchantment

MAGICAL WILL MAY EXERT ITS EFFECTS directly on the universe, or it may use symbols or sigils as intermediaries. The creation of direct effects, like prescience in divination, represents a high point in the art and is just as elusive. Making things happen by either method is referred to as the art of enchanting or the casting of enchantments.

From a magical point of view, it is axiomatic that we have created the world in which we exist. The magician can look around and say, "thus have I willed," or "thus do I perceive," or, more accurately, "thus does my Kia manifest."

It may seem strange to have willed such limiting circumstances, but any form of dualistic manifestation or existence implies limits. If the Kia had willed a different set of limitations, it would have incarnated elsewhere. The tendency of things to continue to exist, even when unobserved, is due to their having their being in chaos. The magician can change something only if he can "match" the chaos which is upholding the normal event. This is the same as becoming one with the source of the event. The magician's will becomes the will of the universe in some particular aspect. It is for this reason that people who witness real magical happenings at close range are sometimes overcome by nausea and may even die. The part of their Kia or life force which was upholding the normal reality is forcibly altered when the abnormal occurs. If this type of magic is attempted with a number of people working in perfect synchronization, it works much better. Conversely, it is even more difficult to perform in front of many persons, all of whom are upholding the ordinary course of events.

In trying to develop the will, the most fatal pitfall is to confuse will with the chauvinism of the ego. Will is not willpower, virility, obstinacy, or hardness. Will is unity of desire.

Will expresses itself best against no resistance when its action passes unnoticed. Only when the mind is in a state of multiple desire do we witness the idiot agonies of willpower. Pitting oneself against various oaths, abstentions, and tests is merely to set up conflicts in the mind. The will always manifests as the victory of the strongest desire, yet the ego reacts with disgust if its chosen desire fails.

The magician therefore seeks unity of desire before he attempts to act. Desires are rearranged before an act, not during it. In all things, he must live like this. As reorganization of belief is the key to liberation, so is reorganization of desire the key to will.

In practice, many difficulties can be gotten around by using various types of sigils. The desire is represented by some pictorial glyph, by a wax image to be wounded, bound, or healed, by the characters of a magical alphabet, or by some image in the mind's eye.

All these serve as a focus for the will. Concentration on these spells should be augmented by some form of gnostic exaltation to cast the enchantment.

When considering any form of enchantment, remember this: it is infinitely easier to manipulate events while they are still embryonic or at their inception. Thus does the magician turn that aspect of chaos which manifests as causality to his advantage, rather than oppose it. The desire then manifests as a convenient, but strange, coincidence, rather than as a startling discontinuity. The will may be strengthened by one other technique aside from the concentrations of magical trance, and that is by luck. The magician should observe the current of his luck in small, inconsequential matters, find the conditions for its success, and try to extend his luck in various small ways.

Those who do their true will are assisted by the momentum of the universe.

LIBER NOX

Initiate Syllabus 2

B
eing the initiate syllabus in black magic, this subject is divided up in accordance with the schema shown in figure 6. We will begin by discussing the spirit of black magic.

Magical power is the key to the heaven-hell of the now. Unaware of this, many slip into the confused, unsatisfactory greyness of small fears and small desires. Then they invent pleasurable or painful hereafters to replace the present. The life force seeks ever the flesh, body, ideas, emotions, experiences, etc., through endless incarnations. For without the flesh, Kia has no mirror for itself, and there is no awareness, no ecstasy, nothing.

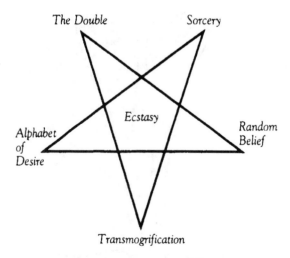

Figure 6. The schema of Liber Nox.

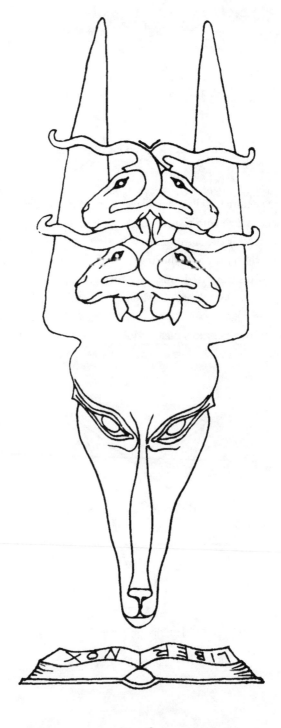

It is either this nothing or the flesh, and the condition of the flesh is dual. Eternal warfare is the price, purpose, and reward of existence.

Kia evolves into myriad experiences, and these are all "self." They are all forms of our awareness. To oppose the great dualities like sex and death, or pain and pleasure, the great evolution of Kia, with lesser fears and desires, is the beginning of failure of satisfaction.

Existence is the great indulgence. Anything less than this, any attempt to avoid part of oneself is to invite loss of form, a self-negation leading to a shrinkage of spirit.

The self alone is God and should recognize itself in all things. For those who uphold limited values bind themselves to mediocrity and failure. Those who self-righteously value their own contradictions are mighty on this earth. Our dual nature is all morality; it is foolish to be other than we are. Acceptance and living without restraint I will call the highest virtue.

The greatest sinners are the greatest saints, though they may be unconscious of this. Great people are greatly dual. And doubleness is not the end of it. Every moment the consortium of "I" puts forward a new face. I am not who I was seconds ago, much less yesterday. Our name is multiple. I am a colony of beings sharing the same envelope. And Kia, the self-love which binds them together, will one day hurl them apart—attempting even death for its satisfaction.

What are gods but humans wielding the force of chaos? To us nothing is true; everything is permitted. There is no purpose to our existence; we are free to choose our own. We have bound ourselves to earth forever and reincarnate at will. For the universe is mad and arbitrary in her ways. Nothing is unchangeable except change itself. The only universal principle is the universal lack of principle. Yet the great goddess Chaos will lend some of her power to those who can become her favorites.

And must our dualities be kept in coexistent separation to retain the power of satisfaction? The small pleasures of the gourmet, semi-digested, overripe, putrefying food have come almost full circle to the consumption of excrement, with little satisfaction. Rather would I take plain wholesome food for the body, and taste the nectars of revulsion and delight but occasionally.

Reserve Kia for works of inspiration, ecstasy, and magic. Seeking pleasure is the surest invocation of displeasure; one falls back rapidly into the general greyness. But if an emotion be pushed to the furthest realms of non-necessitation, then there is nothing to balance ecstasy but ecstasy.

Gnosis is the mechanism by which Kia draws back from the flesh in preparation for the mighty indulgences of magic. A great saving to accomplish a greater spending.

Kia is nascent energy seeking form. It has been called the great desire, the life force, or self-love. It can be represented by Atu 0, the tarot Fool, or the joker. Its heraldic beast is the vulture, for it ever descends to take its satisfaction among the living and the dead.

The experience I received was this, that my innermost self or soul or spirit

Was no thing
formless
without quality
nameless
pure power,
Yet it was anything that it touched,
I am this illusion
and
I am not this illusion
Amen.

Sorcery

SORCERY IS THE ART OF USING material bases to effect magical transformations. The advantage of using such material bases is that the power
residing in them can be built up over a period. Four main types of material
base are used: those which contain reserves of a particular type of power
such as fetishes, talismans, spirit traps, and amulets; those which act to
carry some effect to its target like powders, philtres, wax images, and knotted cords; those which act as a basis to receive divinatory impressions; and
those which act as an anchor for some aetheric form which can be sent like
a magical weapon. In addition, blood and semen may be used as sources of
the life force. Various other bodily secreta and excreta—hair, fingernails,
spittle, etc.—can be used to provide magical links with target persons.

Talismans, amulets, and fetishes are charged by a process analogous to
evocation. Talismans are usually the recipient of some simple charge that
evokes strength, courage, health, virility, no-mind, sleep, or some other
emotion or state of power in its possessor when he concentrates on it. If
the talisman is well enough made, it should continue to evoke its effect
even when it is not being used for concentration. The form and composi-
tion of the material base should be suggestive of the desired effect.

Amulets are objects containing a portion of the aetheric and life force
with a particular task to perform and are semi-sentient. They can be created only by the strongest forms of evocation. They are fashioned in the
form of a small human or creature, and during the evocation an aetheric
duplicate is placed inside the material form. They are most commonly
made to protect places or persons within a short radius of action. When
such things are created by a group or tribe of persons over a long period,
they are known as fetishes.

Any sort of material base is a spirit trap in some way, but some substances, notably crystals, absorb aetheric imprints very readily. Quartz
crystals, which are large and readily available, can be used to pick up
impressions by leaving them near charged places, persons, or objects. If
a spirit or elemental is discovered to be inhabiting some place, it can be
trapped by plunging a crystal into its form.

Enchantment by sorcery is carried out with the aid of various powders,
philtres, concoctions, wax images, and knotted cords. The material base

can be composed of anything suggestive of the desired result and may include possessions or parts of the body of the intended victim. As the material base is being compounded, the magician does everything possible with concentration, visualization, and gnostic exaltation to imprint desire into it. The charged matter is then taken and placed where the victim will come in contact with it.

Instruments of sorcery also find their uses in the mantic art. Most divinatory tools serve only to receive impressions from the operator's magical perception. Charged instruments contain a residuum of formless aetheric energy which actually amplifies the impressions. Most devices in this class are magical mirrors, crystal spheres, highly polished surfaces, and pools of dark liquid or blood. The "mirror of darkness" is an instrument fashioned from black glass or natural obsidian.

It should be carried concealed close to the body. It may be charged by using it as a focus for concentration in the magical trance of no-mindedness. One gazes into it unwaveringly for long periods until it opens like a pit or tunnel beneath one. Only after this aetheric tunnel has developed is the mirror of darkness ready for use. The perception reaches through the tunnel as the will directs it to other regions of time and space.

Magical weapons are created by building an aetheric duplicate of some existing device, such as a wand, sword, dagger, pointed bone, or dart. The aetheric form is kept inside the material base until projected forth by a strong focusing of the will. Skilled sorcerers are able to reach through the mirror of darkness to the target or victim and hurl the magical weapon down after it. Close proximity or even contact with the target is otherwise required.

Personal blood sacrifice may be made to a magical weapon, or it may be made the focus of an orgiastic rite and anointed with sexual fluids. But with or without these adjuncts, it is intense, prolonged concentration which imbues these devices with power.

Amulets and weapons of great power are sometimes given personal names by which they are controlled. Sometimes such devices have acted quite independently of the incompetents into whose hands they occasionally fall.

The Double

THE DOUBLE IS DESCRIBED in all magical traditions from ancient shamanism to the Egyptian "Ka," the "Ki" of occult martial arts to ideas of the soul or ghost, and the modern occult concept of the astral body. It is most commonly seen or experienced when the physical body passes close to death.

It does not have a definite fixed shape, although there is a natural tendency for the life force to hold it in the same image as the physical body. Even when it is exteriorized in the body's image, it is not necessarily visible to ordinary perception. Like all aetheric matter, its effects on ordinary reality are variable and depend on the ability of the life force to make a real effect coalesce at some point. Thus, the double is able to penetrate solid matter, but at other times it may have a degree of tangibility and be able to effect material happenings.

In the occult martial arts, it is projected by visualization just beyond the striking surfaces of one's body, and a sharp yell accompanies the physical blow. A portion of the aetheric force may even be left inside the opponent to cause what is called the delayed death touch. The force may also be projected beyond one's body to inform one of the movement of enemies behind or projected to the surface of one's body to ward off blows. In most forms of psychic healing, this same force is projected, commonly through the hands, onto someone else.

The double may also be made to take on various alternative forms, most commonly animal form. Theriomorphic (beast-like) manifestations of the double are often atavistic. They cause a form of possession reflecting the behavior patterns of the animal concerned. These patterns may lie dormant in our memory, or it may be that we have access to aetheric memories. Whatever their source, these atavisms create terrifying effects. Even if the aetheric beast form is kept within the physical body, it may manifest as strange physical prowess, the ability to cower wild animals and confuse and frighten us humans.

Projected beyond the body, it can serve as a vehicle for the consciousness to experience the mode of travel and abilities of the animal. Skilled magicians may attempt bizarre composite forms like gryphons and basilisks as magical vehicles. In the normal cycle of birth and death, the life force carries little or nothing from one incarnation to the next. The projection of the double is the basis of deliberately carrying things over into a new incarnation at death.

Of all the techniques of gaining access to the double, narcosis is the least controllable and most dangerous. Nevertheless, since time immemorial magicians have been smearing themselves with pastes compounded from the solanaceae alkaloids—thornapple, nightshade, and henbane—and consuming various other hallucinogens and trance-inducing drugs as well, just for this purpose. Visualization is the weakest technique used on its own, but it can act as a model on which to build that peculiar bodily sensation which comes from movements in the aether. Sometimes it is felt as heat, as a sort of itching, or as an ache. One can only persist with imagining something until a sensation develops. Sharp, high-pitched yells and exhalations of breath are used to help project the force in martial arts and occult yogas.

Dreaming is the most challenging and complete method of freeing the double. Ordinary dreams are an ingenious jumble of half-forgotten events, hopes, and worries. They are a more graphic form of the mental chattering, fantasizing, and daydreaming that the waking mind does. In the same way that the day mind learns to differentiate between real things and fantasy, so can the dream consciousness learn the difference between real and fantasy dreams.

Real dreams are the key to the double. The first step in creating the double is to establish it in a real dream. The hands are the most easily visible part of the body. Dream consciousness is particularly linked to the sense of sight. The magician strives to see the hands of his double in dream. The desire to do this is concentrated on intently before sleep every night for as long as it takes success to crown one's work. During this time, the hands may feature in many ordinary, idle dreams which may become very complex and bizarre. This is not the desired result.

Success is an abrupt and discontinuous experience. The command to see one's hands is suddenly remembered as one realizes one is dreaming. Suddenly the hands are there in full clear view. The shock is like being rudely awakened from daydreaming or bursting through a membrane. To prevent the shock causing awakening, the experience should be repeated several times.

Then the magician resolves to see a particular place visited in waking hours also. Summoning the hands in dream, the magician looks at them, then moves them aside and tries to find the place in the material world. The magician must strive to get all the details of the place correct and to be there at the same time of day that the dream takes place. Eventually the double actually is at the desired location. When this much has been attained, there is no limit to what may be eventually achieved. Yet one must be prepared to devote every night for the rest of one's life to developing these powers.

Transmogrification

THE METAMORPHOSIS TO BLACK MAGICAL CONSCIOUSNESS.

Chaos, the life force of the universe, is not human-hearted. Therefore, the wizard cannot be human-hearted when seeking to tap the force of the universe. The magician performs monstrous and arbitrary acts to loosen the hold of human limitations.

The magical life demands the abandonment of comfort, conventionality, security, and safety—for competition, combat, extremes, and adversity are needed to produce higher resolutions and personal evolution. An air of desperation is required in a life lived close to the edge. One must be living by one's wits.

In a stagnant environment the body-mind creates its own adversity—disease and fantasy.

Only in extremes can the spirit discover itself. A fluid environment is required as a vessel for magical consciousness. Only a fluid environment can conform to beliefs about it and be subject to the subtle magic forces. Only in mutable circumstances can divination come into its own.

Therefore, abandon all fixed patterns of residence, employment, relationship, and taste.

Among the titles of Kia is Anon. Anon freely transmogrifies its arbitrary personality, refusing any identity defined by its environment. Residing in the ultimate freedom possible on the plane of illusion, it has choice of duality. Everything which exists for it is a form of desire, for this is the universe in which it willed to incarnate. If this were believed to be either heaven or hell, one would feel free to do anything. It is only the fear that it is neither which imprisons us.

The idea of mind or ego as a fixed attribute or possession of self is illusory. All that can be said of Kia is that the amount of meaning one experiences is proportional to Kia's manifestation in one's circumstances.

Kia is felt as meaningfulness, power, genius, and ecstasy in action. Outside this, nothing is true.

Wizards doeth as they wilt on this illusory plane, knowing that nothing is more important than anything else and that anything done is only a gesture.

Wizards thus become free to do anything as though it mattered to them. Acting without lust of result, they achieve their will.

Figure 7. The sigil of chaos, the eight-rayed star or Octaris, is the prime symbol of chaos magic. It also acts as a device of recognition and as a mirror of darkness for communication between its adepts.

In the arena of Anon compete numerous selves, souls, familiar spirits, demons, obsessions, and an infinity of possible experiences. Each game is short, and then the pieces are hurled through death into unrecognizable new configurations.

Only the style and spirit of Anon's play survive transmogrification, unless the aetheric body has achieved great integration.

The acts of the black magician will bind the magician's Kia to earth forever, but magicians fearful of the loss of their previous occult learning may strongly visualize the sigil of chaos shown in figure 7 at the moment of their death.

Ecstasy

BEING A RESUME OF THE *quadriga sexualis,* a series of somatic trances involving the death posture and rites of Thanateros. The science and art of the sacred alignments between the magical will and certain forms of gnostic exaltation are shown in figure 8.

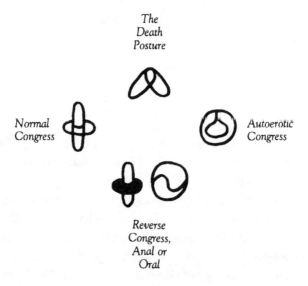

Figure 8. *The quadriga sexualis and their magical uses.*

The pinnacle of excitation and the cave of absolute quiescence are the same place magically and physiologically. In that hidden dimension of one's being hangs the hawk vulture of the self (or Kia), free of desire, yet ready to hurl itself into any experience or act.

The endless variations of the gnosis of quiescence are all the death posture, but the following may be of particular use in part or whole:

Kneeling in the dragon position, hands flat on thighs, spine erect, initiates stare fixedly at the image of their own eyes in a large mirror before them, about two feet away. The temple is best completely featureless, black or white. They may have prepared themselves previously by concentration on one of the magical trances or by some intense effort of convergent thinking.

Gazing at their own eyes, the initiates stop thinking. No amount of "effort" in the usual sense will avail. Infinite patience will barely suffice. The attention must be continually turned to the eye image until the thinking eventually gives up. Any sort of distortion of the image is symptomatic of thinking and to be avoided.

Success is characterized by certain phenomena for which it is useless to strive. There may be a loss of physical perspective or body image. The body may begin to feel vast or microscopic. These phenomena are characteristic of sensory deprivation. They are not the desired result but indicative of a loosening of belief.

The eyes are then closed and the void is entered as completely as possible. Some image may be used as a receptacle of thought if it be not completely annulled. A visualized shape will do. Hopefully this too can be allowed to fade, leaving Kia hovering in an immensity. From this condition of transcended desire, it can hurl itself into any form of magic. Inspiration or atavisms can be dredged up from recondite corners of the awareness by sigils; will and perception can be extended into new dimensions.

If the foregoing proves insufficient to achieve gnosis, stand on tiptoe, eyes closed with arms locked behind, the neck stretched and the back arched, the whole body straining to the limit. The breathing becomes deep and spasmodic as the crucifixion continues. Oblivious to everything except the strain and tension which rack the entire being, attain the void, as this too is suddenly removed and the magician falls exhausted supine to the floor.

At this point, evoke the sensation of laughter. Reflecting on the meaninglessness of anything that comes into consciousness, laugh aimlessly at everything. Receive the grace of being swept up onto the divine madness of ecstatic laughter.

The death posture is chief among the rites of Thanateros, for it can be applied to inspire, reify, and exhaust, as well as transcend, desire.

Normal congress, the genital embrace of persons of opposed sexes, should in any manifestation, magical or not, inspire the participants with something, if only mutual attachment. Employed magically, it can provide amplified inspiration for almost anything.

The lunar current of the priestess is observed for a day on which the libido is strong. In the priest, the libido is strengthened by conservation. Retiring to the place of working, each attains the silence of the mind by any favored method. For example, they may kneel before each other in the dragon position and perform the preliminaries of the death posture. Then they become conjoined in a stimulatory embrace. As congress occurs, the subject for inspiration or its sigil is meditated on as the mountain of excitation is gradually climbed. As the final step is taken, the desire is abandoned to the subconscious. After the cataclysm, the celebrants aim to remain vacuous but alert, to allow the inspiration to manifest.

The reification, or making real, of a desire is possible through the autoerotic mode of the quadriga sexualis. In this, the mind is kept blank while the sexuality is ignited and brought to a pitch by touch alone. The body is in a supine position, eyes closed, with all other senses annulled as far as possible. As the initiate climbs the mountain of excitation the mind must reject all images and fantasies. As the body goes into the orgasm phase, and in the seconds following, the whole force of the will and perception is focused on the desire or, more conveniently, its sigil. In that brief instant when one is no more, the alignment is made, the obsession formed, the demon born, or the sigil charged, enchantment sent forth.

Eroto-comatose lucidity is a variant of this form of congress in which the desire is to reach the borderland state between consciousness and unconsciousness through which the subconscious images and divinatory impressions flow. The sexuality is stimulated again and again and, if need be, again and again and again, until the consciousness slips into the shadow world. In practice, the body should be neither too tired nor too comfortable, so preventing the unconsciousness of deep sleep. The body may be unable to sustain many orgasms, particularly if male. Karezza, excitation stopping short of orgasm but repeatedly approaching it, may be employed.

Exhaustion of desire is a magical process which works on the principle that wished-for events so often seem to occur after we have forgotten about them consciously. It is because the life force then acts through the aetheric tensions we first created, but of which we are no longer aware. The chosen desire is concentrated on through all phases of the arousal, discharge, and aftermath of these forms of congress for as long as it takes until the mind begins to react against or get sick of it.

The conscious desire, rather than any sigilized form of it, should be used. As soon as the reaction against the desire begins to manifest, the whole thing is banished from the mind by a forcible turning of the attention to other matters. Conscious desire and reaction annulled, the desire will manifest at some time in the future.

Random Belief

VARIOUS STAGES IN THE BELIEF CYCLE of the self are provided in the following sections. Try each or any of them for a week, a month, or a year. This exercise may save one an unnecessary incarnation or two. It may also help to make clear the aeonic mechanism which creates the various psychic millennia of past and future history. The beliefs are given in order, with number 1 understood to follow on from number 6 in a circle. Atheism and chaoism are presented in both their early and degenerate phases to make clear the stages of change and to permit the use of the sacred cube.

Dice Option Number 1: Paganism

The gods show themselves in all things. In the elements, tempestuous and placid by turns; in the seas, the mountains, the green fields, in the hail and in the lightning. They show themselves as various animals, and they show themselves in metals and in stones. Most of all, they show themselves in the minds of humans impelling them to love, to war, to fortune, or to disaster. The gods watch over everything in the world; there is no thing not under the auspices of some god or other.

For in all things there is both substance and essence. The gods came out of chaos, and from the gods came the essences of all things—some gods giving essences to some things and others to different things. Humans contain the essences of all the gods.

What is good or what is ill is what is pleasing or displeasing to the gods. But what is pleasing to Mars may not please Venus. Hence there is war in heaven even as there is war on earth. Yet by making an appropriate invocation or offering, we may set matters aright and gain their favors. If we live always in devotion to our patron god and do not displease the others overmuch, our shade will go at death to rejoin the essence of its deity.

Dice Option Number 2: Monotheism

There is but One God who created everything.
He created man in his own image.
He gave man free will to do good or evil.
Good is what pleases God, evil displeases him.

After you die, God will reward or punish you
For pleasing or displeasing him.
God also created angels and demons.
These are spirits with free will;
Some remained good, some became evil.
These spirits help man to become good
Or tempt him to do evil.
If you stop doing something evil,
God will be pleased.
If you stop doing something you enjoy
For God's sake, he is also pleased.
You may pray to God and ask him for help.
You may worship him with prayer also.
By this he will be pleased.
To know how to be good and please God
You must obey the teachings
And the authority of the religious hierarchy
He has established on earth
As the one true religion.

Dice Option Number 3: Atheism

The idea of God or a personal soul is a hypothesis we have no need of. Besides, there is not the slightest scrap of material evidence that will stand up to examination. Let's stick to what's real, shall we?

There is always some sort of a reason or explanation for everything, even if we haven't managed to work it all out yet. But we're doing pretty well. I mean, you've only to look around yourself—the whole universe works on a sensible cause-and-effect basis; it's only hocus pocus if you're too primitive to figure out how it works. Free will, for instance, is probably just an illusion caused by some defect in the neuroelectro-biochemical plumbing in the brain. But we'll all go on using it until we find out. After all, enjoyment is the whole point in life. The only sort of morality or law worth having is that which stops fools from spoiling their own or other people's enjoyment in the long run.

And when you're dead, you're dead.

Until we find evidence to the contrary.

Dice Option Number 4: Nihilism (Late Atheism)

Material causality is everything. Science can probably explain away everything. There is nothing which is not caused by something else.

But this is no explanation.

The world now seems accidental, arbitrary, and without meaning. We can know *how* everything happens but there is no reason *why*. The universe has become predictable but meaningless. That is the burden of intelligence, of being able to see through it all. There is obviously no spirit or personal survival after death. Hence there is no reason to do anything or, for that matter, to restrain from doing anything. Even this is to deceive ourselves, for there is no such thing as free will. One cannot help but get involved in doing because one happens to be. All motivation is just an attempt to put the body-brain in a lower-energy, less tense state, even if by a roundabout route.

There are no absolutes in terms of importance, goodness, meaning, or truth that do not arise from the accidental structure of the body-brain and its surroundings.

We are just living out the chaotically complex forces which spawned us and which will one day reduce us to nothingness again.

Everything we will ever do is just a result of how we are made and what happens to us. For all our pretense of free will, we are an accident running a fixed but unknown course.

Dice Option Number 5: Chaoism

As above, so below.
I am the universe.
The life force in us
Is the life force of the universe.
The subtle force in us (aether)
Is the subtle force of the universe.
The gross matter in us
Is the gross matter of the universe.
To chaos, nothing is true
And everything is permitted,
Though it has limited itself
To the principle of duality
In building this world
for itself.

(For a further elucidation of these beliefs, consult *The Books of Chaos* in their entirety.)

Dice Option Number 6: Superstition (Low Chaoism)

All phenomena having come from the one source, there exist mysterious connections between things with similarities.

All like things contain the same signature or essence; they share the same spirit. This essence or spirit can be made to go into other things by bringing the signature-bearing objects into contact with whatever is being treated. This is the principle of contagion.

All things being connected in diverse, mysterious ways, one can take augury from anything about anything of which it reminds one. There is nothing that is not an omen about something else to those who but have the wit to know it.

And by similar wisdom, anything can be affected by performing the required action on some other thing that reminds one of it. Like attracts like, the principle of similarity.

Wisest of all are those who know the most deeply hidden connections. They are able to be reminded of the obscure by the more obscure. They know what sacrifices are to be made to adjust or placate the essences of things. Morality is the avoidance of misfortune. One's next incarnation will be as whatever creature was most reminiscent of one's activity in the previous life.

The Alphabet of Desire

EXCEPT FOR THE CURIOUS CONDITION of laughter, which is its own opposite, emotion follows a dual pattern—love and hate, fear and desire, and so on. The following "alphabet of desire" includes all the basic root emotions arranged as complementary dualisms in a form suggestive of the classical gods, or Ruach of Kabbala.

Pagan philosophers saw human qualities mirrored in nature and cast these giant reflections of themselves as gods. It is therefore unsurprising that most pagan cosmologies contain a complete spectrum of our psychology in god form.

The main divisions of emotion have been equated with planetary god forms. Each of these principles manifesto in three important forms, represented here by the alchemical principles of

Mercury   Sulphur   and Salt or Earth

The mercurial (exalting, spiritual) form indicates the cathartic, ecstatic, gnostic mode. Overstimulation of any emotive function creates a mental paroxysm in which the whole consciousness may be caught up. This is experienced as a great release or catharsis and, at higher levels, ecstasy. Finally, the one-pointed consciousness essential to mysticism and magic may supervene, in which case the life force can act directly. The gnostic condition is also the key to radical changes of belief or conversion. Any belief presented in this condition is likely to be retained due to the hyper suggestibility of the vacuous state of the mind.

The sulphurous (quickening, active) form indicates the ordinary basic drives to copulate, to destroy, to be attracted by favorable stimuli and repelled by harmful ones. This is the normal functional mode of the emotion from which the ecstatic and earthly modes are derived.

The earthy (heavy, sluggish) form is evoked when an emotion is baulked of expression or becomes tainted with an admixture of its opposite. It turns in on itself rather than seek fulfillment in action or ecstasy.

Table 2. Emotional Duality

Coagula	Solve
The principle of attraction, coming together	The principle of repulsion, separation, avoidance
Sex ☽	♄ Death
Love ♀	♂ Hate
Desire ♃	☿ Fear
Pleasure ▽	△ Pain
Elation ⧊	⧎ Depression

The greater duality which rules all emotions is shown in table 2. Figure 9 on page 68 shows us that the root of every emotion is always its opposite.

Armed with this self-knowledge, the magician may ever ride the shark of his desire across the ocean of the dual principle to a gratuitous ecstasy. Anticipating the earthy dysfunctional modes, he may transmute their energies and obtain his satisfaction in other forms. (If an alternative desire or its sigil is made the focus of concentration in the climate of a negative emotion, it will soon be realized.)

The twenty-one principles have each been given a simple pictorial glyph and a one-word mnemonic. The glyphs can be employed in various spells and sigils, but the words are mostly quite inaccurate attempts to capture a feeling. The twenty-one principles can be equated with the trumps of the tarot if desired. Kia is equated with the Fool.

There is additionally a supplementary alphabet of four principles to cover the somatic emotions of the pain/pleasure and depression/elation dualities.

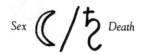

Sex ☽/♄ Death

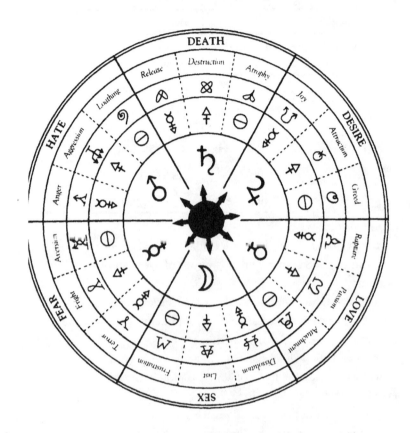

Figure 9. *The root of every emotion is always its opposite.*

The force which creates is also that which destroys. The cellular mechanisms which make reproduction and growth possible are also those which cause ageing and death.

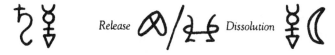

(Glyphs: escape from the dual condition, implosive conjunction)

The death posture includes all trances intended to bring the mind to complete stillness. Concentration on a single stimulus, a thought, an image, a sight, or a sound may hasten the effect of removing all other stimuli. At the moment of the most profound and utter stillness, the magician controls the universe.

Sexuality most often brings a fleeting glimpse of ecstasy. The magician first observes chastity for a period, then takes every measure to attain the highest point of excitation. Ordinary lust is transcended, having served its purpose; the consciousness soars to new peaks of excitation and may pass into something else altogether.

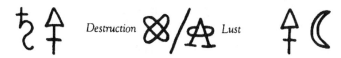

(Glyphs: antagonism, sexual conjunction)

Lust, the impulse to seek sexual union, is a necessary function of the organism and remarkable only in the staggering variety of fetish objects to which it can be directed. Bloodlust and the urge to wanton destructiveness serve few useful purposes. Their existence is inexplicable except in dual terms. The desire for union with various things and persons is co-existent with an equally strong desire for separation from various phenomena. In extreme forms, this manifests as the desire to lay waste to certain aspects of one's universe with a frenzy which parodies sexual lust.

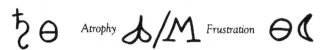

(Glyphs: loss of form by dissolving, failure of conjunction)

As frustration is baulked lust, so are the boredom, laziness, depression, and self-disgust of atrophy a failure to destroy or separate oneself from undesirable events. The attempt to capitalize one's lust or destructiveness by indulgence for entertainment is also a sure evocation of frustration and atrophy.

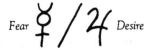

Fear / _Desire_

May it not be that our wished-for treasure islands lie precisely within those images of horror and revulsion we normally reject?

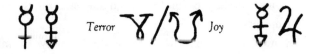

Terror / _Joy_

(Glyphs: black pit of fear, upward, outward leaping)

Fear, if pushed to a high pitch rapidly, will paralyze the mind. Strange alternative magical perceptions may then sometimes be glimpsed; and the will, if focused on a single objective, is mighty. Terror as a tool of magic is an essential ingredient of many initiatory and mystic schools.

The gnosis of joy is more difficult to attain, but a flight of geese against a sunset, contemplation of religious imagery, or intense nostalgia have, for some, been enough to tip the scales of mystic perception.

Fright / _Attraction_

(Glyphs: being attacked, coming together)

The normal reactions to threatening or inviting stimuli, such commonplace feelings require little comment save that in a civilization such as ours, far removed from natural phenomena, almost all fears and desires are socially induced or imaginary.

Aversion ... *Greed*

(Glyphs: inescapable unpleasantness, engulfing)

As a sage once observed, desire is the cause of sorrow. These are important dualities for "civilized" society. Aversion designates the anguish, misery, sorrow, grief, or embarrassment of being unable to separate oneself from phenomena of fear because of past desire. Conversely, greed is the condition of being unable to satisfy desire because of past fears. Daily do our greed and fear assume grotesque and bizarre proportions.

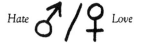

Hate / *Love*

Seeking to avoid violence, we have made a virtue of suppression of anger. This damages the emotional constitution. Denying oneself anger, one loses all the rapture of love. Be ye people of great passions.

Anger / *Rapture*

(Glyphs: transcending rage, flame of passion)

Raging anger is rarely violent and never effectively violent. Ask any warrior. Giving vent to mild anger is cathartic; the head is cleared of tensions, and the body relaxes. Raving blind rage is a magical state of mind. It is useful for casting one's will upon the universe and may, for skilled practitioners, be a gateway to trance states.

Entrancement is also a feature of rapture. Bhakti yoga, the way of love of a god, has parallels in Western mysticism. The force of all-consuming love may carry one into the power of the mystic void.

Aggression / *Passion*

(Glyphs: a weapon, embracing)

Passionate devotion to one's mate, offspring, and tribe is as natural as the impulse to give battle to thieves, enemies, competitors, and predators. Now that society's aggression has been institutionalized on the national scale, it has to be ritualized on the personal and regional scale as sport.

(Glyphs: anger turned inward, useless appendage)

The condition of loathing results from being unable to forget, avoid, or destroy an object of hate. That is, one is unable to break one's attachment to it. Attachment itself is a form of love in which the loved becomes a mere useless appendage when passion is baulked of fulfillment, and a partial reaction, an element of loathing, has entered in.

(Glyph: concealed at the center of the mandala)

Rejected by science, which cannot explain it away; derided by religion, whose piety it deflates; used only to embarrass pretention in art and philosophy—verily it is a tool of magic. In the ecstatic laughter of humans, I see their volition toward release.

(Glyph: downward flashing lightning splitting asunder)

Shattering of one's expectations is the device of all the best humor. The archbishop loudly farts; the free energy of our destroyed beliefs manifests as laughter. This function is protective. If we did not laugh at our broken expectation, we would go mad eventually. By the amoral cultivation of laughter, the magician can shrug off all losses and avoid entering averse states altogether if desired. Crying is an infantile form of deconceptualization, laughter, designed to protect the eyes and summon assistance.

Conceptualization

(Glyph: a receptacle for conception)

Another form of wit, the pun, depends on forging a connection between two ideas. An infinite series of weak jokes of this kind relies on liberating the free energy of surprise when the penny drops. The "aha!" or "eureka!" reaction of discovery is the same emotion, and its enjoyment is what impels people to meddle with sciences, Kabbala, and even crossword puzzles. This functional mode impels us to thought and discovery; it is the motivation for intellection.

 Union

(Glyph: double upward flashing lightnings of cancellation)

The ecstatic laughter of divine madness is the sweeping up of every perception into a vortex of surprise at its very existence. Everything is suddenly and amazingly not as it was. Yet simultaneously it seems more exactly as it was than before!? Conceptualization and deconceptualization occur simultaneously. Language descends inevitably to the paradoxical as one is swept up into the ecstasy.

Such usually accidental paroxysms may be cultivated by forms of the death posture and by willful evocation of laughter on encountering all things.

Some emotional states depend more on the activation of the purely physiological rather than the psychological responses. These are dealt with in a supplementary alphabet. The supplementary alphabet of the somatic emotions (shown in figure 10 on page 74) corresponds to the four elements: earth, air, fire, and water.

No emotion is a purely mental event; all emotions are dependent on complex chemical and nervous reactions to environment. The somatic (body) emotions, however, may be recognized as having a more direct relationship to the senses and general tone of the nervous system. The ecstatic functional and negative modes of each state will be discussed

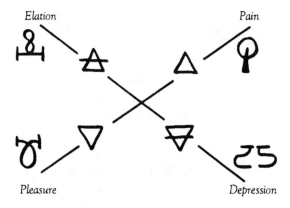

Figure 10. Supplementary alphabet in Malkuth, the somatic emotions.

under general headings. The somatic emotions are intimately associated with the larger alphabet. Depression is linked with many of the earthy modes, and pain or pleasure with most of the functional modes. These linkages are two-way, in that stimulation of one may evoke the other and vice versa.

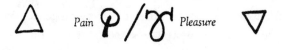

(Glyphs: penetration, gentle touch)

Instinctive movements toward or away from stimuli of various types and intensity are linked with the emotions of pleasure and pain. The range of such instincts is very limited—attraction toward softness, warmth, and bland tastes; repulsion from injury, temperature extremes, and acrid tastes.

Overstimulation of these emotions leads to ecstatic states. This is difficult to achieve for pleasure, but pain has magical application in rites of initiation, purification, and sacrifice.

Complete agony is ecstasy.

Hedonism, the search for gratuitous pleasure, inevitably leads to an epicureanism of pain. The hedonist rapidly falls back into indulging progressively more poisonous, revolting, and debilitating pleasures to provoke a reaction from his exhausted senses. Hedonism and masochism exhaust themselves uselessly into the numbed greyness of dulled faculties.

(Glyphs: sagging, rising upon itself)

The general condition of the nervous system depends on health and the emotions going on within it. Emotions of the ecstatic mode enliven the system and cause a general elation. Those tending toward the earthy model of baulked or frustrated emotions cause a general depression.

Such feelings are generally called happiness or misery. The "dark night of the soul" which follows after certain forms of mystic exaltation is simply the general tone of the neuroendocrine system swinging wildly from elation to depression.

Mere ideas follow suit.

By the alphabet of desire is explained our "inability to make progress in emotional terms." We are ever confined by the dualities of pleasure/pain or happiness/misery, no matter how ingeniously we manipulate our environment.

Lament not those who suffer war, fear, pain, and death, for these are but the inevitable accompaniment to love, desire, pleasure, and sex. Only laughter can be gotten away with for free. Some have sought to avoid suffering by avoiding desire. Thus, they have only small desires and small sufferings, poor fools. The wise seek satisfaction in that which repels as well as that which attracts. Plunging into experience thus, we may be partakers of the dual ecstasy forever and ever. (See figure 11 on page 76.)

Even if satisfaction of some emotion be baulked, and we are unable to rise above it to ecstasy, then it is possible to transmute the trapped energy for other purposes. Whatever it is that we were emotional about should be forgotten, and another desire, magical or mundane, should be substituted for it. Even the desire for laughter can be substituted. It will soon be realized.

There is no escape from the cycle of desire, but armed with such knowledge, a measure of freedom of desire may be won. The alphabet of desire may be called the arena of Anon, for when Kia attains complete anonymity—freedom from identification—it may wander the alphabet as it wills.

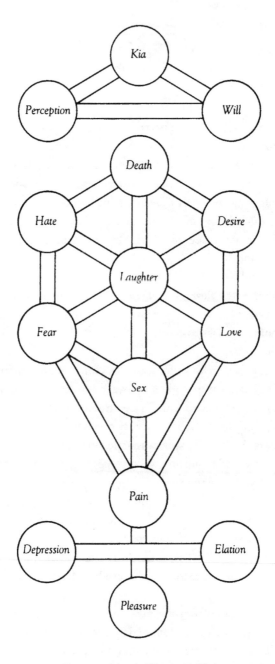

Figure 11. This shall be my Kabbala.

The Millennium

HUMANITY HAS EVOLVED through four major states of consciousness, or aeons, and a fifth is on the horizon. The first aeon comes out of the mists of time. It was an age of animism, shamanism, and magic when the rulers of humans had a firm grasp of the psychic forces. Such forces conferred a high survival value on puny naked people living in intimate communion with the dangers of a hostile environment. This form of consciousness has left its mark in the various underground traditions of witchcraft and sorcery. It has also survived in the hands of several aboriginal cultures in which the powers were used to enforce social conformity.

The second pagan aeon arose with a more settled way of life as agriculture and city dwelling began. As more complex forms of thought arose and people moved further away from nature, the knowledge of psychic forces became confused. Gods, spirits, and superstition uneasily filled the gaps created by loss of natural knowledge and mankind's expanding awareness of its own mind.

The third, or monotheistic, aeon arose inside the pagan civilizations and swept their old form of consciousness away. The experiment was begun once in Egypt but failed. It really came into its own with Zoroastrianism, Judaism, and later with Christianity and Islam. In the East, Buddhism was the form it took. In the monotheistic aeon, people worshipped a singular, idealized form of themselves.

The atheistic aeon arose within Western monotheistic cultures and began to spread throughout the world, although the process is far from complete. It is far from being a mere negation of monotheistic ideas. It contains the radical and positive notions that the universe can be understood and manipulated by careful observation of the behavior of material things. The existence of spiritual beings is considered to be a question without any real meaning. People look toward their emotional experience as the only ground of meaning.

Now some cultures have remained in one aeon while others have swept forward, but most have never completely freed themselves of the residues of the past. Thus, sorcery tainted pagan civilizations and even our own. Paganism taints Catholicism and Protestantism. The time required for a leading culture to break through into a new aeon shortens as history progresses. The atheistic aeon began several hundred years ago. The monotheistic aeon began two and a half to three thousand years ago. The pagan aeon began about six thousand years ago with the beginnings of civilization, while the first animistic and shamanistic aeon goes back to the dawn of humanity.

There are signs that the fifth aeon is developing exactly where it might be expected—within leading sections of the foremost atheistic cultures.

The evolution of consciousness is cyclic in the form of an upward spiral. The fifth aeon represents a return to the consciousness of the first aeon but in a higher form.

Chaoist philosophy will again become a dominant intellectual and moral force. Psychic powers will increasingly be looked to for solutions to humanity's problems. A series of general and specific prophecies may be extrapolated from current trends to show how this will come about, and what role the Illuminati will play in it.

Decades, possibly centuries, of warfare lie ahead. The remnants of monotheism are collapsing fast, despite the odd revival, before secular humanism and consumerism. The technological, atheist superstates are trying for a stranglehold on human consciousness. We are entering a phase which may become as oppressive to the spirit as medieval monotheism. The production/consumption equation is becoming increasingly difficult to grasp or balance as the consumer religion of the masses begins to dictate politics and degrade the environment.

More and more mechanisms for the forceful regulation of behavior have to be introduced as population density pushes individuals to seek ever more bizarre forms of satisfaction in material sensationalism. The problem with any belief system is its tenacity and inertia once it is established and dominant. The medieval religions murdered millions to protect their own hegemony. Innumerable crusades, jihads, burnings, and massacres were committed. In the end, no level of persecution could stave off the inevitable ascendency of atheism.

Now it is the atheist superstates which are supplying the arms and dropping the bombs in support of the hegemony of consumer capitalism or consumer communism. And this is only the beginning. The blind logic of technology and consumerism will cause alienation, disaffection, greed, and identity crises to rise to such catastrophic levels that the situation may explode into a very destructive war. There may be a breakdown of society which may take the form of an antitechnological jihad. This will not resolve the contradictions of the system but merely introduce a new dark age and slow the changes down. However momentous these events may seem, if they happen, they will not affect the movement of consciousness in the long run. They will only affect its timing. But the Illuminati must be ready to exploit the changes which will definitely occur. Among these are:

The Death of Spirituality. Fixed ideas about the essential spirit or nature of humans will be completely abandoned as an "emotional technology" becomes more refined. Drugs, obscure sexualities, faddism, strange entertainments, and material sensationalism are a preliminary groping toward this end. Chemicals, electronics, and surgery will only tend to enslave. Gnosis, the alphabet of desire, and other magical methods tend to liberate.

The Death of Superstition. Prejudice against the possibility of the occult or supernatural will give way in the face of a developing "magical technology." Telepathy, telekinesis, mind influence, hypnosis, fascination, and charisma will be systematically examined, refined, and exploited as methods of control. We may see magicians working behind barbed wire and in underground cells also.

The Death of Identity. Ideas about a person's place in society, roles, lifestyle, and ego qualities will lose their hold as the cohesive forces in society disintegrate. Subcultural values will proliferate to such a bewildering extent that a whole new class of professionals will arise to control them. Such a "transmutation technology" will deal in fashions, in ways of being. Lifestyle consultants will become the new priests of our civilizations. They will be the new magicians.

The Death of Belief. We will abandon all fixed ideas about what is absolute or valuable and what constitutes morality as a "psychological technology" develops. Techniques of belief and behavior modification in the military, in psychiatry, in places of detention, in propaganda, in the schools, and in the media will become so sophisticated that truth will become a matter of who creates it. Reality will become magical.

The Death of Ideology. Ideas about what form human aspiration should take will give way to a science of the preservation of the control mechanism—government and its agencies. These may become global or semi-global, but their primary concern will become preservation of the government, for or against the people. Primitive cybernetics will mushroom into a "political technology." Governments will be provided with the choice of either accommodating themselves to coordinating proliferating human variety or seeking to reduce that variety by repressive measures.

LIBER AOM

The Work of the Adept

The rituals of the degree of the adept are secret, yet they are stated here in the most explicit form that language permits. Only by perfecting oneself in the work of the initiate can one attain the empowerment required to use the techniques of the adept. Anything less than this leads to failure, disaster, and death. The methods are given here that some sight of the eventual goal of the work might be glimpsed.

In the work of the adept, the aspirant has risen above the use of all symbolic systems save reality itself. The playthings of the initiate—sigils, gods, demons, and the instruments of the sorcerer—become reabsorbed or retained only for teaching purposes. The need for magical weapons will be restricted to the ceremonial. Tools will be direct prescience and

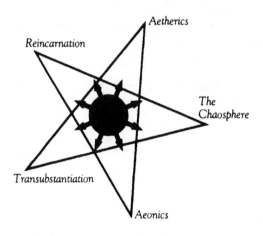

Figure 12. *The schema of Liber AOM.*

enchantment. In the lower grades, knowledge and ecstasy were taken as a guide to progress. For the adept, only the ability to work magic is used to measure the increasing strength of "spirituality." During the work of transmutation to magical consciousness, the adept is completed and becomes one who liveth chaos in perpetuity. The force may take the adept anywhere, perhaps to project the requisite personality dualities and gather persons to form autonomous satraps of the Illuminati, or perhaps simply to withdraw manifestation from the plane of duality entirely and cease to exist.

For out of chaos arise the two prime forces of existence, the solve et coagula of existence. The light power and the dark. The light power is the expanding, outgoing, dualizing, increasing expression of chaos, responsible for the new birth, creation, incarnation, and variety. The dark power is the contracting, returning, transcending, withdrawing expression of chaos, responsible for death, dissolution, reabsorption, simplicity, and return to the source.

These twin forces lie at the root of all mystic quests and all forms of magical and mundane action. They are the basic spiritual principles of the universe. Adherents of one will always call the other black. Thus, the expansion into existence can as easily be called the dark ascent into matter, and the withdrawal from existence can be called the return to the light. But this is mere moral philosophy and ultimately devoid of meaning. The positive and negative paths of magic converge and form a unity which itself diverges from the mystic paths. While mysticism is at the root, concerned with following either the light power or the dark, magic aims to play one off against the other. The magician aims to become a center of creation and destruction, a living manifestation of the chaos force within the realm of duality, a complete microcosm, a god.

The work of the adept is arranged under the headings shown in figure 12 on page 81.

Aetherics

THE AETHERS IN THEIR VARIOUS DEGREES of density are the means by which magic and miracles are affected. The adept may require these abilities to accomplish his will on earth, or he may simply wish to develop them as a test of the mystic abilities.

Operations of this kind may be classified according to the density and origin of the aetheric forces employed. Operation of the dense aether is used to accomplish gross feats such as levitation and the short-range occult martial arts. The force responsible for this resides in the belly just above the navel and may be extruded as lines of force to any part of the body and to locations outside the body's surface to a distance of several meters. Levitation (which includes the ability to walk on water and fire, as well as in the air, or at fantastic rates across the earth) is accomplished by supporting the body's weight with these aetheric force lines. In the case of fire walking, the force is used to repel the heat and flames.

To get these force lines to come from the body, the adept sits cross-legged on the ground and attempts to jump like a frog from this awkward position. The exercise is usually conducted in darkness and the jumps attempted with the lungs fully inflated. Within as little as three years' practice, one may begin to learn how to exert the aetheric force against the ground. A cushion may be used to protect the tissues of the posterior in the meantime.

Movement of the force within the body is easier to achieve and most commonly is approached by trying to develop a warmth or itching sensation at various points. The adept should be able to achieve freedom from the attack of virtually any form of disease and ensure longevity by directing the psychic force to any site of weakness. If the force can be drawn to the hands, then it may be projected just beyond for healing purposes or to deal fatal blows to an enemy.

The dense aetheric force is the basis of most superhuman feats: the abilities of leaping to great height, clinging to smooth vertical surfaces, the deflection of weapons, and the strength of madness.

Operation of the subtle aethers involves the transference of thoughts and emotions by the agency of the aethers which normally link the life force to the brain. The simpler the message and the greater its emotional charge, the more easily is it projected. As with all things, destructive effects are much more easily created than constructive ones. The adept may begin to strengthen abilities by selecting a small specimen of plant life and trying to make it wilt and wither by the force of unaided will. As it begins to succumb, the adept may reverse the current of his will and resuscitate it again. They may practice telepathic contact with various beasts, dogs being particularly suitable for this purpose. At odd moments they may pick various persons around themselves and make them stand up, sit down, or move around and do particular things.

One of the classical tests of psychic ability is the power of casting an aura of subjective invisibility about oneself. This does not involve any alteration in one's optical properties, but surrounding persons are somehow prevented from noticing one's presence. The complementary ability to this is the projection of an aura of charisma or fascination, which may have application in the formation of sundry cults and messianic sects. These powers arise from a constant, semiconscious, intense concentration on the characteristics one wishes to project.

Transubstantiation

The penultimate metamorphosis,
the achievement
of constant magical consciousness

In all things know that
I AM THIS ILLUSION

In all things know also that
I AM NOT THIS ILLUSION

But there is no thing more ineffectual
than to be always
only half in the world and half out of it,

Thus, at the moment of one's doing
Be In It
Identify Completely
Live It
And in the interstices between all one's doing

Reside in the Void
Be Vacuous
Have No Mind

Know that by this one may attain
to complete freedom
from the consequences of one's actions.

The Chaosphere

THE CHAOSPHERE IS THE PRIME radiant or magic lamp of the adept—a psychic singularity which emitteth the brilliant darkness. It is a purposely created crack in the fabric of reality through which the stuff of chaos enters our dimension. Alternatively, it may be considered a demonstration of the axiom that belief has the power to structure reality.

The chaosphere may be given a material form which acts as an anchor to locate its chaotic and aetheric manifestations. The shape shown in figure 13 is only one of a number of possibilities. It consists of a sphere with eight arrows radiant directed toward the vertices of a cube. Thus, to the thinking mind, it may be said to variously represent a perspective sculpture of the four axes of the geometrically impossible hypercube or the two interpenetrate tetrahedra of the light and dark forces. Such twists of illogic may be useful in the creation of an essentially paradoxical object.

Figure 13. *The chaosphere.*

It is colored the deepest black, for this is to give it all colors simultaneously and to provide it with the greatest potential for emission and absorption. The central sphere is hollow to permit the inclusion of various objects, and one of the arrows is detachable as a magical weapon. However, the physical shape can take any form that the ingenium of the adept suggests; it is a trifling matter compared to the psychic preparation which goes into its construction.

The chaosphere is charged and opened into the magical dimension by filling it with aetheric life force which has been paradox modulated.

The life force may be supplied by any method over which the adept has mastery—blood sacrifice, sexual emissions, projection of the body's aetheric force of transference by concentration during ecstatic gnostic rites, or by other methods. The paradox modulation is achieved by imprinting the life force with all manner of contradiction and impossibility. Any two opposite and mutually exclusive ideas or images can be simultaneously employed—the imaginary metaphysical principles of fire and water, or Nuit and Hadit, incompatible geometrics, simultaneous blackness and whiteness, mathematical zero and infinity. Any manifestation of paradox and possibility can be used, and many different ones should be made to serve.

If the adept be operating a temple, initiates and students may assist in formulating the chaosphere, and it may then function as a god or fetish. The more power that is put into it, the wider will be the breach opened into the chaos dimension and the greater will be the free chaos energy liberated.

Once it is made, it will begin to supply raw chaos force to anyone coming into physical proximity with it. For this reason, the uninitiated should be kept well away, lest they become deranged. The sphere also exists as a vortex or door through which the magical will and perception can reach easily to the other regions of existence in the manner of a powerful magical mirror. The erection of operating chaospheres at various points about the earth will tend to hasten the immanentization of the eschaton, the change of aeon.

Aeonics

THE FIFTH MAGICAL AEON EXISTS only in embryonic form. Its precise manifestation hangs in the balance. The new Chaoist aeon may develop into an Aquarian Age or a time of totalitarian tyranny. (See "The Millennium.") As agents of the fifth aeon Illuminati, adepts may occasionally break their meditation to alter the course of history. For existence is the chariot of chaos, and humans the most refined vehicles available. Adepts may be concerned that the society of humans continues to evolve ever better forms to support the divine incarnation of Kia on earth.

They may work to build the new aeon's technologies—the technologies of emotion, belief, magic, transmutation, and politics—which will free people from spirituality, prejudice, superstition, identity, and ideology.

And they may seek to create cults, orders, covens, and cabals to spread the secret wisdom and to destroy the satraps of old aeon spirituality and its thought forms.

For the fifth aeon has the potential to be a great renaissance when mighty deeds will be done upon the earth, and Chaoist philosophy shall hurl humanity to the corners of the galaxy and into the very center of its being.

Either that or a new dark age.

Reincarnation

INTEGRAL REINCARNATION IS THE FINAL metamorphosis. It is a supreme ritual of sex and death by which adepts can attain the degree of master.

In the ordinary course of events, the memory, personal aether, and life force of Kia disintegrate as the material body disintegrates. New beings are built up from the pool of the universal life force in the same way that they are built up from the universal pool of matter. The personal life force then finds its way back into incarnation, broken up into millions of parts scattered into many beings.

Adept magicians, however, can so strengthen their "selfs" by magic that it becomes possible for them to carry those "selfs" over whole into a new body. In an exceptional display of power, it may even be possible to retain specific memories in aether form. It is by such a process that adepts can achieve the final degree of mastership. The supreme rituals of sex and death exist in three forms: the red, black, and white rites.

The Red Rite

Adepts may attempt to reincarnate by blood, in the sense that they strive to create exact duplicates of themselves in their offspring. Cruelty begets cruelty far more efficiently than genius begets genius, yet many aristocracies have tried to circumvent the chaotic reshuffling of characteristics that reproduction entails by inbreeding and strict education. The teaching of ideas and beliefs often proves ineffective in the absence of the experiences which led to the adoption of those ideas and beliefs. Thus, the faster the world changes, the weaker the hereditary route to immortality, of sorts, becomes.

The Black Rite

This rite consists of the forceful entry of a spirit into the body of an already inhabited being. It is both dangerous and unreliable and used only in certain peculiar and desperate situations. It will sometimes result in there being a double life force in one body, if the invading spirit fails to displace the occupant. In this case, the victim will seem to have gone mad. In any case, the memory of the taken-over being remains, and the invading being will have to work through this. In the darker forms of this abominable rite, the adept may attempt to so dominate a victim that the victim effectively becomes a copy of the adept while the adept still lives.

The White Rite

In this rite, an integral reincarnation is achieved, but the receiving body is a fetus randomly selected by the escaping spirit in astral flight. In the event that one's friends, followers, and disciples might be unable to locate one's new manifestation and ensure that it receives a proper magical education, it may be wise to carry an aetheric marker. At death, some sigil emblematic of one's magical aspiration is fiercely visualized. It may later be apparent to clairvoyant perception in the aetheric constitution of the developing infant, or it may serve to cause some recognition or affinity with the symbol if it is perchance accidentally stumbled upon in the new existence.

PSYCHONAUT

A Manual of the Theory
and Practice of Magic

Introduction

After some centuries of neglect, advanced minds are turning their attention to magic once more. It used to be said that magic was what we had before science was properly organized. It now seems that magic is where science is actually heading. Enlightened anthropology has grudgingly admitted that beneath all the ritual and mumbo jumbo of so-called primitive cultures, there exists a very real and awesome power that cannot be explained away. Higher physics now suggests that the universe runs on something more akin to sorcery than clockwork.

The magic art itself is undergoing a profound renaissance. More than enough Kabbalistic and goetic trivia has been harvested from the hallowed British Museum reading room. In this new aeon, the thrust of magical endeavor is toward making the actual experimental techniques work regardless of their religious or symbolic associations. The techniques of magic will be the hypersciences of the future. The origin of these arts lies not in medieval or even pagan civilizations, but is to be found in its most developed form in animistic and shamanistic cultures. Before history began, humanity knew a strange and terrible power which has gradually slipped from its grasp. Humanity now stands on the brink of rediscovering that power. This is a book of source material and a work of reference for those who seek to perform group magic or to work as shamanic priests to the community. It is a companion volume to *Liber Null,* which constitutes a manual of the individual sorcery of Thanateros.

The sections of this book may be read in any order; it is an encyclopedia of related essays. Ideas which are not fully explained in one section will be clarified in the section from which they originate.

NEW AEON MAGIC

A n ancient Chinese curse runs, "May you live in interesting times." Since the fall of the Roman Empire, there have rarely been more interesting times than these. Whenever history becomes unstable and destinies hang in the balance, then magicians and messiahs appear everywhere. Our own civilization has moved into an epoch of permanent crisis and upheaval, and we are beset by a plague of wizards. They serve a historic purpose, for whenever a society undergoes radical change, alternative spiritualities proliferate, and from among these a culture will select its new worldview.

It is for the wizards to determine how that new spirituality will manifest. Most will end up crying in the wilderness or being put to the stake in various ways, but a few will bequeath to humanity a greater gift than they realize. Orthodox hierarchical monotheistic religions are a spent force, spiritually and intellectually, although there will be some bloody battles before they are completely finished. Science has brought us power and ideas but not the wisdom or responsibility to handle them.

The next great advance that humanity will make will be into the psychic domain. There are many encouraging signs that this is beginning to occur. In this new field of endeavor, we shall rediscover much of the magical knowledge that the ancient animists and shamans once possessed. Of course, we shall know it under different guises and will eventually expand on their knowledge immensely.

There is a twofold aspect to the most important magical work now being done. Foremost is an experimental investigation into the actual techniques as opposed to the mere symbolism of magic. The methods of magic are remarkably uniform throughout history and across cultures. It is time to unearth them and make them work. Secondly, it is essential that what might be called a spirituality of magic is evolved. Magic must have its own flavor, its own worldview, and its own philosophy. There has always been a tendency to regard magic as an antique art. All shamanic systems consider themselves to be the repository of only a fraction of the power and knowledge that their traditions once held.

It seems that, in the past, reality was more chaotic and susceptible to magic. Even astrophysics and biology support the mythical view. To look at the furthest objects in the heavens is to catch a glimpse of some of the earliest events in this universe. Here cataclysmic events of unbelievable violence and strangeness are occurring. The fossil record shows that our own planet once shook beneath the feet of immense and totally improbable dragons. It does seem that as the universe ages, matter becomes more ordered and sensible and that the force of magic diminishes. This certainly seems true for the relationship between magic and matter. Except for the occasional metal-bending maverick and the odd shaman who still insists on walking over pits of fire, the grosser magical power seems to be receding.

This is, however, only half the story. A profound change has occurred, and the magic force is now manifesting with increasing strength on the psychic levels. The creativity of consciousness has mushroomed so enormously that the totality of human ideas seems to double with each decade. Science has not caused this; science is one of its many side effects, as are parallel explosions in art, music, and general creativity. On the magical level, the psychic powers are becoming much more accessible. Telepathy, clairvoyance, and astral travel were once won only at great cost by an elect few through extreme measures. Now they are within reach of anyone armed with moderate determination.

The beginnings of the new psychic awareness have acquired a definite subversive flavor. Magic is aligning itself against oppressive forms of order in many fields. Magic is opposed to a psychiatry and medicine designed to patch up the damaged automaton and plug it back into the system. Instead, it would rather that individuals learn to handle their own mental self-defense and treat their bodies with gentler remedies such as herbs.

Magic rejects politics as no more than some people's perverse desire to dominate others. It does well to dissociate itself from this monkey squabble and advocates instead personal enlightenment and emancipation, which are the only real safeguards to freedom. Magic is anti-ideological because the main products of ideological solutions are repression and corpses.

Magic is profoundly opposed to religion. Although a religion may appear benign when it is in decline, at least half of the madness and violent deaths of history have been caused by mindless adherence to religions. Magic is also opposed to the superstition that the world is wholly material and that human actions are not intimately interwoven with the psychic sphere.

To oppose repressive forms of order, which often impose themselves by evil means, magic aligns itself to a vision of chaotic good. Magic's commitment to the good is reflected in its concern with individual freedom and consciousness and its interest in all other life forms on this planet. At the highest level, this manifests as some unspecifiable feeling for the "vibes" generated by human thought and action.

The chaotic aspect of new aeon magic is psychological anarchy. It is a species of operation mindfuck applied to ourselves as much as the world. The aim is to produce inspiration and enlightenment through disordering our belief structures. Humor, random belief, counter-information, and disinformation are its techniques.

To take an innocuous example, I usually advocate astrology persuasively to ordinary people but ridicule it to my magician friends. Humor and random belief allow the use of astrology to disorder what people think either way. Does this mean that I am: (a) lying, (b) mad, (c) enlightened, (d) aware of our ability to live almost any truth?

GROUP MAGICAL EXPERIMENTS

The purpose of structuring group activity with ritual is to generate more power than individual efforts might achieve. Synergistic effects will come into play in a properly synchronized working, and the collective power will exceed the sum of individual powers participating. Group working also makes possible many experiments requiring more than one operator and allows for a division of labor when some participants can contribute abilities which others lack. Group magical work can be performed as a training exercise, as research, or as a procedure for creating effects.

No technique is absolutely fail-safe, and so all activities are experimental to a degree. Some exercises may be so totally experimental in concept and execution that it is not possible to say why we are doing them. If it were possible to say why the research was being done, then it would be unnecessary to do it.

Specific effects will be the aim of most training exercises and techniques, and it is with these that this chapter is mainly concerned. Four areas of experimentation will be examined: psychism, ritual, trance, and dream.

Aside from synchronizing group activity, ritual also acts as a solemnizing factor. The assumption of ritual dress and paraphernalia serves to mark the transition from ordinary activity to something of importance. A uniform has a further function—that of depersonalization. It helps to reduce the importance of individual personality factors and allows the wearers to relate to one another as functionaries of some principle beyond personal considerations. A full-length black robe with hood is most excellent for this purpose, as is nudity. A blank, featureless mask completes the effect to total anonymity.

Group experiments with simple psychism fall mainly into the category of training exercises in telepathy. Several significant considerations apply here. Telepathy is made more effective by the synchronized projection of many operatives to a single target, but a single projector will be more successful than a poorly synchronized mob. Attempts to project or receive a whole string of images to obtain a statistical result is far less effectual than projecting a single image powerfully. Intense momentary concentration on a sigil or symbol is usually best. With repeated attempts, the projecting and receiving capacities become confused and distracted.

The so-called astral doorway experiments in which a single operator attempts to divine the symbolic import of a particular image, sigil, or symbol are little more than exercises in creative imagination performed in a light trance. If the experiment is performed simultaneously by several participants, it can become a basis for the exchange of telepathic imagery. Psychic activities can be synchronized by a ritual in which previously arranged signals cue in particular meditations. The group can concentrate on a symbol or mantra, or they can be led through imagery suggested aloud by a leader.

Many other methods of raising power will also act to synchronize the participants in a ritual. The essence of magical dance is that it should be a frenetic circumambulation of a circle about a fixed point, accompanied by collective mantrical invocations. Balance is maintained by keeping the gaze fixed on the center of the circle. Over-breathing, flagellation, and stimulants can also be employed to heighten excitement and gnosis. Sexual arousal is difficult to control and coordinate except on an autoerotic basis and is more often a feature of the aftermath of an excitatory ritual than a means of stimulating controlled group gnosis.

Full rituals of the meditative or ecstatic variety are usually directed toward one of four objectives: enchantment—making things happen directly by magic; evocation—making things happen through the agency of various demons and elementals; invocation—the summoning of various entities and thought forms for the inspiration of their knowledge and conversation; and divination—obtaining knowledge by direct magical means.

The simplest way to orchestrate these rituals for group work is for a presiding officer to perform the main ritual sequence and have the participants deliver their visualizations, mantras, ritual movements, and invocations on certain prearranged cues. To be effective, the ritual must work like an automatic mechanism in which power can manifest without distraction or hesitation.

A variety of trance states from mild suggestibility to deep hypnosis can be used for group magical work. One operator will persuade one or more subjects into a receptive condition by suggestion or invocation, the subject being in a relaxed or lightly drugged condition. The psychic censor is less active in the trance state but often acts to block the awareness of magical events from reaching other levels. When the trance state is controlled by another person, this problem can be overcome. The trance candidate can be directed to seek information clairvoyantly and to relate it to the operator. Conversely, the subject's will can be directed to perform a magical act that the censor would normally prohibit.

One danger with trance experiments is that the operator's influence over the subject may gradually extend itself to nontrance states as well. Another is that the memory and imagination can become very active in trance states and begin to delude both operator and subject. For these reasons, trance experiments should be performed infrequently and for objective results only.

It is possible to perform group magical experiments on the dream level. The main difficulty with working in dreams is to make the command to act in a particular way on the dream level penetrate to the dream state. Some form of prearranged ritual incantation or visualization can serve to impress the desire to act magically into the deeper levels of the mind before sleep.

The so-called astral sabbat is the main type of magical experiment performed on the dream level. The participants arrange to dream of being present in each other's company at some real place with which they are familiar. The participants can either enter sleep in separate locations at a prearranged time or sleep together in one place. In the second case, it may be possible for the first persons to exteriorize in their dream bodies and attempt to astrally awaken the others. The initial purpose of such sabbats is to achieve a common perception. In subsequent experiments, acts of will can be attempted. Sexual attraction can be used to supply a motivating force to meet on the dream level, and "flying ointments"—unguents that can induce a sense of flying—can assist with the exteriorization process.

LEVELS OF
CONSCIOUSNESS

Since psychology began, people have never tired of devising new ways of compartmentalizing the mind. All these schemes are more or less arbitrary and rarely related to observable structures within the brain. Many schemes merely reflect the moralistic prejudices of those who devised them. Basically, all schemes fail because the complexity of the mind exceeds the sophistication of the scheme. Even such apparently basic divisions into conscious and subconscious are questionable. The entire contents of the mind seem to be subconscious; it's just a matter of recall, and there is a complete scale from the easily accessible to the inaccessible with no reason to draw an arbitrary line at any particular point. Most of what is described as the realm of higher consciousness seems to be a mixture of wishful moralistic thinking and a few of the more obscure and dysfunctional instincts and drives.

Neither psychology nor psychiatry has made much headway with attempts to understand how the contents of the mind interact. The cause and cure of madness remain as obscure as ever. Whatever the relationship between the contents of consciousness may be, it is evident that consciousness occurs on a scale of five states, herewith: gnosis, awareness, robotic, dreaming, unconsciousness.

Unconsciousness has few uses beyond allowing the body to rest and keeping the organism out of harm's way during those hours of darkness for which it is not adapted.

Dreaming, which usually, although not invariably, occurs in sleep, has many functions; it allows the mind to digest conscious experience and make emotional adjustments to it. It also provides a window into the psychic dimension and into the less accessible regions of memory.

The robotic state allows us to perform all the automatic behaviors that waking life demands: walking, eating, driving vehicles, and all the million other little meaningless tasks which require no thought once they have been learned.

Awareness occurs when the mind produces some nonautomatic response to a stimulus. Some minds will be provoked into awareness only by unusual outside events; other minds may be able to self-stimulate themselves into awareness. The degree and duration of awareness that any stimulus provokes may vary from very little to a lot, depending basically on intelligence.

The level of gnosis occurs when the mind becomes intensely conscious of anything. This is not the same as thinking intently about something. Gnosis is a state of intense awareness in which thinking ceases and the object of consciousness holds the attention of the mind completely. Terror, anger, orgasm, and various quiescent meditations will provoke this condition.

The robotic level has come in for a lot of schtick from mystics generally. While it is useful to be able to drive a car or walk automatically, it is obviously undesirable to live one's whole life like this. Nevertheless, the robotic level has many other uses. It is into the robotic level that inspiration or clairvoyant impressions often intrude, and it is partly from the robotic level that enchantments are cast—rather than from the awareness level. Most of the world's geniuses have had some sort of robotic hobby or distraction, which they use to create a vacuum in their awareness into which something useful could manifest. Similarly, most methodological forms of divination are designed to occupy the mind with a thoughtless robotic task. Also, when casting enchantments, it is essential that the actual procedure can be performed without having to think about it.

Although the awareness level may be the forum in which we refine our methods and theories and experience many of our most meaningful moments, it is of very little use in performing magic. In fact, the more someone is centered in the awareness level, the more difficult magic may be for them.

The gnostic level is the very font of magical powers and mystic states of consciousness. Despite the tidal wave of verbose nonsense that mystical experience provokes on the intellectual plane, it is quite simple to state exactly what gnosis is and how to reach it. Gnosis is intense consciousness of something, including the ideas of self or nothingness. Most extremes of emotionality (and not just the nice emotions) can initiate it, and so can a profound act of single-pointed concentration on something. This intense consciousness leads mystics into three common errors. It may create the illusion that oneself and the object of concentration are the same thing. It may lead to the conviction that oneself no longer exists, and it may lead to the obsession that the object of concentration is the supreme thing in the universe.

Magically, gnosis is the state which most easily allows the will and perception to reach out and touch realities beyond the mind. The contents of gnosis are far less interesting than what can be done with it. It is of course possible for some activity to be occurring at more than one level of consciousness. The robotic level, for example, continues to function during all but the most arresting moments of awareness, and parts of it still operate even in gnosis. Most trance and hypnotic states seem to fall somewhere between the robotic and dream levels. I have my suspicions that parts of the dream level are operating without our noticing them when we are awake, much as the stars continue to shine in the daytime without our noticing them.

Most people would identify with their robotic or awareness levels, a few artists and madmen might feel most at home in dreaming, and the mystic would locate his real being in the gnostic level. According to the magical perspective, none of these is true. The self is no more than the point at which the formless life force (or Kia) touches experience. Because consciousness occurs only at the Kia/mind interface, we are unable to get at the root of self with ideas. To fill this gap or vacuum we erect an ego. The ego is an image of self and Kia that we build up out of habit. The Kia should be able to find expression at any level and be equally at home in all or none of these states.

Magical training is designed to open up the neglected dream level, to provoke an examination of the contents of the robotic level, and to add new programs to it. It should also teach the method of turning awareness on or off at will, and of entering the gnostic level and acting within it.

The normal human life is spent oscillating between the unconscious and robotic levels, punctuated by odd moments of dreaming and awareness. The magician may well strive to establish a new oscillation between dreaming and awareness with occasional excursions into the robotic and gnostic levels for specific purposes.

MAGICAL COMBAT

The combat of witch doctors and sorcerers occurs either as a result of unresolvable conflicts of professional interest, or else as training exercises or tests of supremacy.

If both protagonists are equally skilled, the results are unlikely to be fatal. Combat between magicians and ordinary people, each with their own techniques and weapons, is likely to be as dangerous to either party as combat between ordinary humans.

Magical combat is to be undertaken with the same seriousness given to considerations of assault, the infliction of distress and disease, grievous bodily harm, and murder. The protagonist who is psychologically unprepared to do these things physically will not accomplish them psychically. Of all possible motives, revenge is the most pointless except as a demonstration and warning to others. Violence is a very blunt instrument and a little reflection may indicate more effective forms of psychic intervention, such as spells of ligature and binding, or operations to change one's adversaries' opinions.

Magical attack takes two forms. At long range, telepathic information is sent which makes the target destroy itself. To make someone fall under a vehicle is not impossible; to make a vehicle fall on top of someone is something else entirely. At short range, it is possible to injure or drain an adversary's energy field using one's own. This demands close proximity, usually contact. Magical close combat of this type is not affected by mere will or visualization, but by projecting a force that can actually be felt, usually through the hands.

More rarely, the force can be projected through the voice or the eyes or carried on the breath. The force originates in the navel area and is aroused by the disciplines of breath, concentration, visualization, and by sexual disciplines. A part of this force is put into the enemy's body to cause a disruption of the vital energies leading to disease and death. The only defense consists of evading contact or in having sufficient control over one's internal energies to be able to neutralize the effects of the incoming disruptive energy.

Psychic vampirism may be an entirely passive and nondeliberate phenomenon, as when young persons live intimately with much older people. Vital energies cannot easily be drawn from a weaker person into a stronger sorcerer unless the sorcerer first kills or severely weakens the victim at close range.

Long-range magical combat depends on projecting self-destructive impulses telepathically. A number of methods exist for avoiding the dangers inherent in this technique. Foremost among them is getting one's apprentices to do the dirty work. The image of the target wounded in the required manner is used to send the attack. Wax images, photographs, hair, or nail parings help to form a connection between the visualized image and the target. To focus the sorcerer's psychic energy, the attack is launched from a state of deepest concentration or from a pinnacle of ecstatic excitement. Hate and anger aroused during a full ritual destruction of the image may serve. The magician may inflict pain on himself, imagining it originates from an adversary in order to arouse fury. A longer method calling for protracted concentration is the "black fast," in which the psychic energies aroused by fasting are directed with malefic intent at the target.

The "death fetish" is a composite method of attack which can be used at any range. The sorcerer compounds a device to carry a death wish to the enemy. Foul and necrotic ingredients, together with something to represent the foe, are ritually prepared with full magical concentration during which the sorcerer adds psychic force for proximity transmission. The fetish is then placed where the intended victim will come into contact with it. A skilled sorcerer may succeed in projecting a purely aetheric entity across space to harass opponents. Magical attack is usually done by stealth. There is very little point in betraying one's intentions, unless the victim is of a highly nervous, paranoid, or superstitious disposition.

The main difficulty with defense from magical attack is that the very act of trying to divine the enemy's precise intention increases one's vulnerability to it. A third party is most useful here. A counterattack on its own is a high-risk strategy if the enemy has already taken the initiative. Riskiest of all is to send back an identical attack. The preparation of an attack inevitably involves generating self-destructive impulses for projection. There is always the risk that this can backfire and doubly so in this case. The situation is analogous to a duel with grenades.

The most effective defenses are provided by sentient or semi-sentient entities. Prolonged obsessive religious activity will, for ordinary people, create a protective aetheric thought form. This effect is partly transferable and explains the difficulty of attacking popular public figures. It is noticeable that when such a figure falls from favor and is stripped of the protective thoughts of followers, then sickness and death often follow quickly. The sorcerer will create entities with more deliberation and care. Entities anchored in talismans, amulets, and fetishes are made by concentrating psychic energies in various objects—sometimes aided by sacrifices of blood or sexual secretions.

In all forms of actual or suspected magical attack, paranoia can be the worst enemy. It is the height of unwisdom to enter situations in which conflict is the only option left. Magical attack is the direct opposite of occult healing, though it uses similar forces. As with all things, constructive activities are a far greater challenge to our skills than destructive ones.

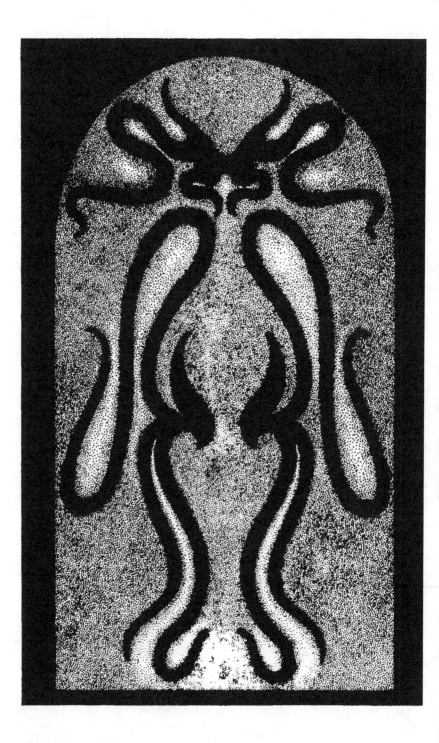

THE RITES OF CHAOS

F ive rites based on the principles of gnostic new aeon shamanism are presented here to cover most situations that the priests of chaos may encounter. They are the "Mass of Chaos," "initiation," "exorcism," "extreme unction," and "ordination."

The Mass of Chaos is a general rite which may be performed for the purposes of invocation, evocation, enchantment, or consecration. The rite of initiation gives the general method for admitting candidates as initiates into inner orders. The rite of exorcism is applicable to psychic infestations of persons, places, or objects. The rite of extreme unction (or last rites) may be applied to the dead or dying body of creatures of all species including our own.

No rites are given for the creation of adepts or masters, for each seeker must devise his own entrance into these grades and await peer recognition. An outline of the requirements for the attainment of the status of occult priesthood (ordination) is given, together with a rite to complete the process.

Each of the rites is given in a general form for adaptation as circumstance demands. Each may be performed by a single operator or by an unlimited number of participants and assistants.

The Mass of Chaos

THIS RITE MAY BE PERFORMED as a sacrament of invocation to raise a particular manifestation of energy for inspiration, divination, or communion with particular domains of consciousness. It may be performed as an act of enchantment in which spells are projected to modify physical reality. It may also be performed to consecrate magical instruments or evoke entities for later use. The rite consists of a minimum of six parts: preparation, statement of intent, invocation of chaos, invocation of Baphomet, oath, and closing.

The preparation will include the making ready of the site, the erection of circles and triangles, the placing of instruments and weapons, and the administration of any chemical or botanical elixirs that may be employed to heighten gnosis. Banishing rituals, meditation, circle dances, and other forms of preparatory gnosis may be used to prepare the participants.

The statement of intent must be formulated as simply, forcefully, and precisely as possible. Holding aloft any material basis that is to be used in the rite, the officiating priest intones the words, "It is our will _____," adding whatever the aim of the rite is to be. The material basis may be some foodstuff for subsequent consecration and consumption. It may be a sigil with which to cast an enchantment or a talisman, amulet, or fetish for consecration. In the event that the basis is to be a sexual elixir, then the priest or priestess stands empty-handed, for the sacrifice is to be of their own bodies.

The invocation of chaos is affected by a barbarous incantation delivered in conjunction with gnostic methods of the operator's choice. The supreme animadversion to chaos is given below, together with a translation which is as accurate as possible within the primitive logic structure of the English language. Drawing the sigil of chaos in the air above the circle and assisted by the visualizations of the same by his assistants, the priest begins:

OL SONUF VAROSAGAI GOHU
I   Reign   Over You   Saith

VOUINA   VABZIR DE TEHOM QUADMONAH
The Dragon Eagle   of the Primal Chaos

ZIR   ILE   IAIDA   DAYES PRAF   ELILA
I Am the First   the Highest   That Live in   the First Aether

ZIRDO   KIAFI   CAOSAGO   MOSPELEH   TELOCH
I Am   the Terror   of the Earth   the Horns   of Death

PANPIRA   MALPIRGAY   CAOSAGI
Pouring Down   the Fires of Life   On the Earth

ZAZAS   ZAZAS   NASATANATA   ZAZAS
(This last line cannot be translated.)

The eight-rayed star of chaos radiant is visualized above the circle throughout, and sacrifices of incenses, blood, or sexual elixirs may be made.

For the invocation of Baphomet, the priest or priestess who is to assume the manifestation of Baphomet attires and visualizes themself in the traditional god form of this power source. Baphomet, as the representation of the terrestrial life-current, appears as a horned theriomorphic deity of androgynous, winged, reptilian, mammalian, and human aspect. The priest arouses from within a resurgence of the chi, or kundalini, or sacred firesnake, as it is variously known.

Other participants may assist by delivering such incantations as the incomparable "Hymn to Pan" by projecting a visualization of the averse pentagram into the priest and, if need be, by administering the *osculum infame*. (This so-called obscene kiss to the devil's hindquarters has been much misunderstood. All that is required is that one breathes onto the peritoneum, the space between the genitals and the anus—inside of which the kundalini awaits to be aroused.) The priest then completes the invocation with the aeonic litany.

In the first aeon, I was the Great Spirit
In the second aeon, Men knew me as the Horned God,
Pangenitor Panphage
In the third aeon, I was the Dark one, the Devil
In the fourth aeon, Men know me not, for I am the Hidden One
In this new aeon, I appear before you as Baphomet

The God before all gods who shall endure to the end of the Earth.

The priest as Baphomet now takes any material basis being used as a focus for the rite and consecrates it to the purpose of the ritual by whatever means the god wills, perhaps by speech, by gesture, or by some unexpected means. The Oath marks the culmination of the rite. Holding the material basis aloft, the priest and all participants affirm:

"This Is My Will"

If the basis be a sacrament, it is then consumed. If it is a sigil, it is destroyed or hidden, while a consecrated object is then covered and concealed for later use.

The closing may necessitate an exorcism of the priest if the trance or possession be deep. Any Baphometic symbols and paraphernalia are removed, and an upright pentagram is drawn over the priest. A full facial lustration with cold water is administered, and he is called forth by his ordinary name until he responds.

A final banishing ritual closes the rite.

Initiation

INITIATION CAN NEVER BE performed according to some set formula. No two candidates will have exactly the same requirements, abilities, and shortcomings. Any order which attempts to initiate by a set formula is displaying a remarkable lack of perception and imagination. Existence itself can be seen as a continuous initiation punctuated by periodic death and rebirth, which have great initiatory potential themselves.

Beyond a certain level, the magician will begin to seek to devise personalized initiatory experiences deliberately, or may feel that something inside begins to intuit these experiences. There is no fixed route which one can traverse to become an adept or master automatically. There are too many variables in existence to make a simple equation possible.

When a student or neophyte approaches the guardians of an organized body of occult knowledge, they will be required to undergo some form of initiation if they survive an initial period of instruction and assessment. An operation of this type should never be repeated at a so-called higher level. If the adepts of an order are not able to completely satisfy themselves with the candidate and take him fully into their confidence, then they have no business initiating. Orders within orders exist only to manufacture hierarchy for its own sake.

The formal initiation will contain all of the following elements:

An Ordeal—to test fortitude and devotion to the order, and to test various abilities that the order may require.

An Empowerment—with certain secrets, powers, and knowledge that are the property of the order.

An Induction into the Order—which may place certain obligations of duty and secrecy upon the candidate.

The Unexpected—the order should be capable of engineering some event which will greatly surprise the candidate, overturn expectations, and force a thought or act which is completely at variance to normal behavior. The practical joke is in many ways a secular survival of initiation rites and a reflection of the general cosmic joke being played out continuously in all our particular existences. In past aeons, this experience was usually provided by some sort of simulated death and rebirth. There are many variants of it—hypnosis, hallucinogens, or temporary reduction to total vulnerability can be used. Terror, physical privation, or ecstatic excitement leading to collapse should be employed with a certain prudence, however.

Ordeals may also be set to test specific abilities, such as power of divination. The candidates might be asked to give the history of a certain object with which they are presented, or alternatively be asked to conjure a pentagram with sufficient force that it can be seen by others present.

While it would be neither desirable nor possible to give exact initiatory formulae, a resume of some of the general methods is presented below. Firstly, some examples of ordeals:

The Initiatory Journey. The candidate is sent (or led) on a journey, perhaps through a forest at night, or even across a crowded city. At various points, the candidate is met and challenged by various guardians and also by persons masquerading as ordinary people. Each will make some demand which the candidate will have to satisfy before being passed to the next point.

Guarding a Station. The candidate is assigned to a location that must not be quit on pain of failure. It might be a tree in a wood or even a lamppost in a public place. Various experiences which force confrontation with fears and desires are arranged to tempt the candidate from the post.

Magical Defense. A circle is drawn on the ground around the candidate. All other officers of the order are outside the circle. No person may cross the circle, and no object which is a physical weapon may transit the circle. There are no other rules. Combat ceases if the candidate submits or when the presiding officer is satisfied.

Secondly, some examples of empowerments: the candidate may be prepared by fasting, by meditation, and, if need be, by various elixirs and then be shown certain forces and entities that the officers of the order may conjure. Alternatively, the candidate may be placed in trance and led through a series of visions. Group rituals designed to produce heightened or altered states of consciousness may also be performed. The candidate may be taught certain techniques or have some instrument consecrated for personal use.

The inductive phase of an initiation informs the candidate of the order's requirements pertaining to secrecy. Blood, nail parings, spittle, and the measure (a cord of the candidate's precise head-to-toe length) may be given in pledge.

Of what might constitute the unexpected part of an initiation ritual I shall say no more. Whereas the majority of the rite is a test of the order's magical and organizational abilities, engineering the unexpected is a test of the order's creativity.

Exorcism

EXORCISM IS OF TWO TYPES: the exorcism of places or objects and the exorcism of persons. A possible third type, the exorcism of animals, is rarely encounted and seldom worth attempting, it being extraordinarily difficult anyway.

The exorcism of persons does not invariably presuppose that some external entity has taken over the victim's mind. We are quite capable of manufacturing our own demons through bad mental habits or in response to peculiar forms of stress. It is worth attempting exorcism as a cure for madness only if the patient specifically complains of some sort of infestation by a seemingly independent entity. Additionally, it is likely to succeed only if the possession began comparatively recently. As the witch doctors say, a person with a bad soul, a long-term mad case, cannot usually be helped.

The conventional religious form of exorcism seeks to replace one obsession with a larger and more powerful one. This involves invoking a god to cast out the demon. This can only work if the candidate has been brought up to fear or revere a god. The exorcist must invoke, and to some extent actually personify, the candidate's god using all the symbolic words, actions, and paraphernalia that go with it. Then the exorcist must command the candidate to abandon the obsession, if necessary by quite forceful behavior. It may be useful if the exorcist makes a show of actually removing the demon. Some form of symbolic trickery is often used at this point.

Psychiatric methods of head doctoring are still incredibly primitive. Most depend on "carrot or stick" techniques. Carrot therapy depends on being as nice and reasonable as possible to the patient and is sometimes effective over a long period. Stick therapy derives from the medieval practice of flogging demons out of people. In these civilized days, it is usually administered with electricity, the scalpel, or the hypodermic. Its effectiveness is questionable.

The techniques of freestyle shamanism offer an alternative to the religious and psychiatric methods. Instead of invoking the candidate's god, the exorcist invokes and personifies the candidate's demons. This approach may be particularly useful with a nonreligious candidate.

After observing the candidate for some time, the exorcist will arrange a situation of complete dominance over the candidate. Then the exorcist takes the candidate through a tour of their private hell. Foul and acrid incenses may be burned; weird flashing illumination and smoke can be employed. The exorcist behaves in a weird and threatening manner, throwing back at the candidate all the peculiarities they have exhibited themselves. Effectively, the exorcist terrorizes the candidate back to normality by showing them how far down the slope they have slipped and how much further they might eventually go.

Gnostic techniques, that is, the generation of intense mental excitatory ecstasy or alternatively extreme meditative quiescence, are effective in the exorcism of persons. In either of these two conditions, the mind becomes hypersuggestible, which explains their use in brainwashing. Many forms of possession have a sexual or traumatic component. The mental energy associated with sexual arousal or traumatic experiences can often become diverted to feed an obsession until it grows into an independent entity or attracts an independent entity. Indeed, this is one of the easiest methods by which the magician can breed personalized familiars, elementals, and demons. It is only by returning to a similar level of mental exultation by various physiological means that such forms of possession and obsession can be challenged and banished. The rite then consists of bringing the candidate through a great catharsis, during which consciousness can be reprogrammed to reject the offending entity which has been built up.

The exorcism of places and objects is of two types. The first is a sham performed in the presence of persons who require exorcism themselves but who are unable to come to terms with this fact. The exorcist is likely to be summoned to situations where it is obvious that the problem lies in the inhabitants of the place and not the place itself as they protest. In this case, the exorcism is performed in their presence with them as the actual target. The exorcist will have to make some pretense of dealing with the place itself, and in this may ask the affected persons for assistance. As they have already exteriorized the obsessing or possessing force imaginatively, or sometimes objectively (as with poltergeists, etc.), the next logical step is for them to take control of it.

If there are objective psychic phenomena, such as materializations, noises, sudden temperature falls, or objects hurled about, the exorcist should not immediately conclude the presence of totally extra-human entities. Humans are quite capable of effecting such manifestations while being unaware that they are doing so.

If, however, the magician can satisfy himself that some place or object is infested with some psychic energy or entity, then a great opportunity presents itself. Instead of banishing, entrapment can be considered. In general, spirits can be coerced by anything with a highly ordered, low entropy structure—the focused human will, magnetized iron, crystals, and, to some extent, very pure water being the most commonly employed agents.

The magician will usually begin stalking the entity by providing it with some basis of manifestation. Depending on available clairvoyant abilities, the magician may employ darkness, incense smoke, or hallucinogens to catch some impression of the entity. A subsequent rite of entrapment centers on the use of a spirit trap, of which by far the most effective are crystals. Salt crystals are often used more or less unknowingly in religious rites, but shamans the world over prefer larger, more stable crystals, particularly quartz. Incidentally, ordinary salt is so good at picking up various bits of random low level psychic debris that many witches and mystics refuse to eat it. Consuming raw salt is somewhat injurious to physical health anyway.

Entities may be coerced into crystals by plunging them into the space occupied by a spirit if this can be determined. Coercion by a strongly focused human will, assisted by prods and passes with sharp magnetized iron instruments, can be effective—particularly if the whole operation is concentrated by a ritual.

Banishment may be accomplished by simply forcing the entity to leave, or it may be performed subsequent to entrapment. Although it is very common, water is a very anomalous substance. On the molecular level, it is not entirely homogenous or random but possessed of a certain delicate structure which is very sensitive to heat, radiation, and psychic ambience. It will receive a psychic charge readily but will dissipate it equally readily. Hence, in rites of banishing, infested salt crystals are usually dissolved in water for a few days. Another variation on direct banishing is to charge a sample of water and simply splash it around in the vicinity of the entity for its disintegrating effect.

The magician may encounter entities which are the psychic remains of human dead. If these still possess any degree of coherence, the magician may have recourse to various forms of unction as given in the next section.

Extreme Unction:
The Final Enchantment

THE ORGANIC FORCES WHICH BRING a being into physical existence will, inevitably, remove it from existence at some later time. It is neither possible nor desirable to prevent this from occurring eventually. Death is a precondition of life. Without death, human life would no longer be human. Death may be a great initiation or a random catastrophe in which the fruits of an incarnation are largely wasted.

At death, there are three parts of a being to be considered: its Kia, its aetheric body, and its physical body. The last will degrade at various rates depending on the gruesome superstitions surrounding its disposal in a particular culture. Only religions which are truly afraid of death have contrived the revolting habits of burial in sealed boxes or embalming.

In the usual course of events, the aetheric body will begin to degrade as the physical body begins to disintegrate. This aetheric body, sometimes known as the soul, contains an image of the body and some of its most powerful memories. If death occurs in a highly emotionally charged way, the aetheric may contain a memory of that experience also. The disintegrating aetheric sometimes gives rise to all manner of quasi-religious experiences in the dying; each may briefly visit the heaven or hell of expectation. The aetheric may appear as a ghost, and parts of it may become attached to places or objects or—rarely—other people. In general, though, it usually dissipates into the aetheric background after a few days.

The Kia is destined to be reabsorbed into the life-force pool of this world, which makes itself known to us as Baphomet. To the mystic, this experience is union with God. To the sorcerer, it is being devoured by the devil, and he seeks deliberately to avoid it. The magician, on the other hand, may well give some consideration to whether to preserve individual awareness or not. Absorption in Baphomet explodes the Kia into infinitesimal fragments, out of which new Kias will eventually form to inhabit new beings. By magical means, it is possible to get the Kia to reincarnate whole without losing its integrity. Such a reincarnation will be unconscious, and no memories will be preserved. Other techniques allow the Kia to carry some aetheric with it, in which case at least some of the main lessons and memories from one incarnation may be preserved for the next.

Magicians must decide for themselves what course of action they will take for their personal "souls." When present at, or shortly after, the death of any creature, the magician has the opportunity to assist as psychopomp, the guide of souls through the otherworld.

Instructions and encouragement may be given verbally to the dying, but in the case of their being comatose, or dead, or belonging to other races or species, the magician will have to rely on telepathic visualization alone to put a message across. The essential points, which may be stated in any form the candidate is likely to understand, are these:

Be without fear as the great metamorphosis begins.

Fantastic and terrifying visions are illusory; laugh at them and reject them; they cannot touch you now, go beyond.

You will come to the secret of your being, which may seem as a dazzling brilliance or as an awesome darkness, or as both these things and more.

It is your choice to become one with this source if you so will it.

It is your choice to remain separate if you so will this instead.

Do what thou wilt.

If you would remain separate, then you must seek new life.

In seeking rebirth, seek the emanations of love, vitality, and intelligence; go where there are strength and freedom.

Ordination

A MAGICAL PRIEST, as distinct from an adept, is someone capable of administering the sacraments and rites of initiation, exorcism, extreme unction, and the Mass of Chaos, and of discoursing wisely upon mysticism and magic to whosoever may require these things. Most adepts will be able to function as priests unless they are following a particularly solitary path. Initiates will find that acquiring the powers of a magical priest does much to further their progress toward adepthood.

Ordination is not conferred by some passive seal of approval but is given in recognition of the demonstration of certain abilities to one's peers. Orders will recognize as ordained priests of chaos those who can demonstrate the following:

Administration of the Mass of Chaos for invocation, enchantment, and consecration.

Performance of the exorcism of places and persons to effect.

Administration of extreme unction to some being whose demise must certainly not have been brought about for this purpose.

The design and performance of an initiation, acting as principal officer.

The construction and use of magical weapons.

Psychic ability in enchantment and divination by any preferred method.

The ability to enter at least one state of altered consciousness or gnosis at will.

The ability to discourse widely, wisely, convincingly, and with authority on magical and mystical matters.

These abilities have to be tested over a period of time, but after satisfactory accomplishment, the candidate proceeds to the rite of ordination proper. For this, perform the Mass of Chaos for magical inspiration and to consecrate the priestly instruments. As many fellow officers of the order as possible will be present to add their own powers to the rite.

MAGICAL TIME

The celestial bodies which exert the greatest physical and psychic effects on the earth are the sun and moon. The effects of the other planets are minute by comparison and quite unrelated to fanciful attributions to antique gods. Astrology, like any other body of knowledge, seeks to enlarge itself, but in doing this indiscriminately, it has become hopelessly vague and imprecise. The planets do influence earth, but the effects are indirect, as they affect us by affecting the sun, and most of these effects are immeasurably small.

The moon does not shine by her own light but by the reflected light of the sun. In bouncing off the moon, the sun's radiations undergo a change in properties, and on reaching earth, they cast the familiar eerie silver glow over everything they touch.

Before the days of the massive psychiatric abuse of tranquilizers, mental hospitals would become veritable bedlams at the full moon. Strong moonlight exerts a general psychophysical energizing effect on a variety of plant and animal life whose growth and behavior are influenced by it. There are few things more invigorating than moon bathing, but if the energy is not channeled into something useful, it can tend to intoxicate and derange the moon watcher.

Away from civilization, the female menstrual cycle becomes synchronized with the moon's phases. The passage of blood tends to occur at the dark of the moon but may occur at the full moon. Ovulation will necessarily occur at the opposite phase. At the time of menses, a woman is in her most psychically powerful and clairvoyant phase.

Many authorities consider that the full moon is the time for works of beneficent magic, healing, fertility, and gain, and that the dark of the moon is the time for malignant sorceries. This is only partly true. All magic works somewhat better at the full moon, as there is more psychic energy about. Harmful magic tends to have worse effects on its victims at the dark of the moon, for everyone tends to be at a lower ebb, but conversely the attack has to be made from a position of lower energy. The exception to this, of course, being women who menstruate at this time. For this reason, many religious and magical systems tend to be fearful of the power of women at this time and exclude them from temples or intercourse with men. On the other hand, some arcane and secretive magical orders have encouraged the use of women's extrapsychical powers when menses occurs at the full or dark of the moon for good or ill, respectively. Oral contraceptives now offer a simple method for effecting synchronization with either of the moon's phases as is desired.

On the general theme of the timing of any magical act, it is worth noting that the best time to perform any magic to affect others is at four in the morning local time. This is the time when the body-mind is at its lowest physiological level. It is the time of dream and the time when most people are born and when most will die.

The yearly seasonal cycle in temperate latitudes exerts a considerable psychic effect through the agency of plant and animal life. The yearly rhythms of sex, growth, death, and decay create a corresponding psychic current which can probably explain most sun sign astrology and make various types of magic easier at certain times. Springtime energies assist beneficent works such as healing, growth, love, and fertility, these being performed around May Eve, April 30. Autumnal energies assist such works of necromancy, death, and darkness as may be performed around All Hallows Eve, October 31.

The effect of the season of birth on health and life outcomes in temperate climates seems to far outweigh other astrological considerations, although the effect declines within the protective envelope of modern civilization. Antipodeans may thus justifiably reverse much of the lore of astrology developed in the northern hemisphere.

The quality of the sun's radiation is periodically disturbed by the presence of sunspots. These are intense magnetic vortexes which move across the surface of the sun and tend to appear in large numbers every eleven years. They appear darker because they are cooler than the rest of the surface of this raging thermonuclear furnace. The immense energy fields associated with sunspots are measurable here on earth and often disrupt radio communication. Sunspots have unpredictable effects on earth; sunspot maxima are more often than not associated with upheaval and disaster in the affairs of men. Events often move into a crisis phase and great changes begin.

The picture is further complicated by a magnetic polarity reversal from one cycle to the next, giving a complete cycle of twenty-two years. Magically, a time of sunspot maxima is the time to set great changes in motion while events are at their most sensitive and unstable, and the slightest push may have decisive consequences.

The last two maxima occurred in 1968 and 1979, marking the onset of optimistic and pessimistic currents, respectively. Let us hope that 1990, 2012, 2034, and so on, herald better times on earth. It remains to be proved that there is a correlation between the twenty-two Atu of the tarot and the twenty-two-year cycle. The Fool would presumably represent one node, and the Fortune the second, in each cycle.

A knowledge of astronomical and temporal cycles ought not act as a restrictive influence on magical activity. Rather it should suggest times when such arts may be practiced with more than normal efficiency.

CHEMOGNOSIS

IMPORTANT NOTE: *To use drugs of any kind is to poison the body. The difference between sufficient dose and overdose is so variable as to display the danger inherent in the use of toxic substances. The author has undertaken comprehensive study of the use of many different types of drugs in a controlled scientific way, ensuring manifold safeguards and protections during the experiments. Neither the publisher nor the author wishes to incite any reader to the irresponsible use of toxic substances and would advise against their use. However, to omit a survey of this historically important aspect of the workings of magical technique would have been to jeopardize the integrity of the whole book.*

Chemical agents of natural and manufactured origin have always played a significant role in animism, shamanism, and magic. These substances can make various occult powers more accessible, but none of them confers magical abilities by themselves. There are four factors which control the outcome of experiments with magical drugs: firstly, the physiological effects of the drugs themselves; secondly, the training and abilities of their users; thirdly, any innate magical forces contained in the substances; and fourthly, any outside magical events which may affect the experience.

On the basis of their physiological effects, magically useful drugs can be broken down into three categories. Hallucinogens are substances which enhance perception. Hallucinations, as distinct from superior perceptions, occur when the subject has overdosed or failed to direct perceptions to any purpose, and the experience becomes a disordered trip around the imagination. Disinhibitory agents, such as alcohol and hashish, make it easier to attain the gnostic states of frenzied excitement required in various ecstatic rites. Hypnotic or narcotic substances are those which give rise to various degrees of trance and unconsciousness.

Now most drugs in any of these classes will exhibit all three types of effects at various doses. Small amounts of narcotics are stimulating in many cases, and larger doses may be hallucinogenic. Excessive doses of disinhibitory agents may cause stupor and hallucination. Hallucinogens themselves may be stimulating in small doses but cause trance in larger dosages.

Furthermore, all drugs will cause poisoning, coma, and death at some level of dosage, although this may be only at extreme levels. The training and abilities of the users of the drugs account for many of the differences of effect noticed at lower dosages. Quantities which may evoke only mild euphoria or nausea in untrained subjects may be sufficient to allow the adept to enter trance or ecstatic states. The directing of perception is also essential if one is to commune with magical phenomena rather than just have a pleasant or nauseating time. The directing of perception can be learned in nondrug meditation, or it can be brought about by the presence of an adept, or it can be caused by magical forces contained in the drug substance. Failure to direct perceptions is the cause of all meaningless and horrific drug visions.

There may be innate magical forces in a drug if it is made from a living thing or if it has been prepared specially to contain some occult force. For this reason, botanical drugs should be collected with the utmost care and respect. In return, the spirit of the species may yield its secrets to the user: such knowledge as where to find the plant, what its nature and properties are (curative and otherwise), and a knowledge of other creatures and forces having a relationship to it.

Some preparations may contain certain nondrug elements which have occult properties, such as part of an animal with which the sorcerer is seeking communion. When using a refined or purely chemical substance, it is wise to perform an invocation beforehand. At the very least this will direct one's perception, and it may succeed in placing a magical charge in the substance itself.

Outside events may also serve to direct perception. An experienced initiate can lead the neophyte into the correct visions or demonstrate a particular phenomenon to the neophyte's enhanced perception.

Now, briefly, an exegesis of the magical drugs in common use and their effects: "Flying ointments" are found at a variety of points in magical history and in many cultures. The essential ingredients are a grease base and one or more of the poisonous solanum species (datura, henbane, or deadly nightshade, and sometimes aconite or wolfsbane). The ointment is smeared on the forehead and around the thighs and was occasionally applied internally to the female genitalia using a broom handle, hence the myths. The alkaloids of the solanaceae cause drowsiness and unconsciousness during which hallucinations of flying occur and real astral travel is possible. The aconite alkaloids help in the general numbing of the body. All these alkaloids carry a severe risk of fatal poisoning, however, and it is unwise to overdo or ingest the mixture. With this type of drug, it is preferable to use only sparing amounts, then attempt willed astral travel while asleep rather than comatose.

A wide range of hallucinogens are available to stimulate magical perception. Synthetics such as LSD do not possess any intrinsic magical quality but produce dazzling erratic visions which, although they may be emotionally charged, seem only to reflect the expectations or fears of the user. Because of the fleeting and fantastically distorted nature of LSD experiences, it is notoriously difficult to direct perception of particular visions within it. Whereas, in the early days of its use, LSD carried a certain joyous oceanic vibration, nowadays it seems to have acquired an aura of paranoia and madness.

Although it is probably impossible to direct the trance to magical ends, nitrous oxide gas produces startling visions of an intensely inspirational nature. It often seems that this simple substance taps the very seat of inspiration itself, but the insights it brings have an infuriating tendency to slip through one's fingers on awakening. Nevertheless, it gives an enticing taste of something approaching formless samadhi.

Naturally occurring hallucinogens provide a far richer source of magical perception. *Amanita muscaria*, the fly agaric toadstool having a red cap and white spots, contains a variety of alkaloids including bufotenine. This substance is also found in glands behind the eyes of certain toads, which may explain their use in medieval witch's brews. It is also significant that *Amanita muscaria* bears the name toadstool; indeed, it is virtually the archetypal toadstool in folklore, presumably because of this chemical similarly. No toad has ever been seen sitting on one by choice.

A similar group of hallucinogenic alkaloids exists in species of small Psilocybe mushrooms. Something very strange has happened to this species. There seems to have been no reference to them at all in any folklore outside the Americas until recently, very recently. Although virtually every other psychogenic herb and fungus has been known for centuries,

Psilocybe has remained unknown and catalogued as an uninteresting and seldom found little toadstool. It seems that what we are witnessing here is the sudden proliferation of a virile and hallucinogenic mutant within an otherwise insignificant species. It is to be hoped that after a few years it does not disappear again as mysteriously as it appeared.

The little mushrooms produce all the interesting effects of the *Amanita* but without the unpleasant side effects. They are also highly communicative if approached with respect and will show the seeker many aspects of their collective being as well as offering glimpses into the self and the universe.

With all types of excitatory and trance-inducing drugs, the trick is to use just enough to stimulate the required condition but not so much that one loses control of it. Trance-inducing substances include narcotics like opium, tobacco, or mandrake decoctions, and various anesthetics like ether and chloroform. Excitatory preparations include alcohol, hashish, and small quantities of hallucinogens.

All these substances require an additional ecstatic technique to direct the perception to produce a useful effect. In general, chemical agents are only useful in receptive magic—such as astral travel, divination, and invocation—and after a while, the adept should be able to obtain these experiences without chemical assistance. Chemical agents find very little application in more active forms of magic such as sigil casting and enchantment. In magical combat, their use may prove disastrous.

An afterthought: I would not counsel anyone to tread too deeply into the mire of alchemy, but the "black elixir" of that tradition was almost certainly essence of toad.

NOTE: *All drugs are poisons and the previously mentioned substances are capable of acting as lethal toxins. With many natural psychoactives, the difference between a fatal and a merely psychoactive dosage is impossible to assay by amateur methods. These techniques are mentioned only for the sake of historical completeness.*

MAGICAL
PERSPECTIVES

Physical processes alone will never completely explain the existence of the universe, life, and consciousness. Religious answers are just wishful thinking and wanton fabrication cast as a veil over a bottomless pit of ignorance. To explain their occult and mystical experiences, magicians are forced to develop models beyond the scope of materialistic or religious systems. To the magician, it is self-evident that there is some other level of reality than the purely physical. Medieval magicians thought that their powers emanated from God or the devil. In fact, magic works equally well in any god's name for good, evil, neutral, or indifferent motives. Whatever the nature of the other reality, there is obviously no need—beyond the psychological—to anthropomorphize it.

Many scientific disciplines begin by not observing any sort of vital spark or consciousness in material events and proceed to deny that these things exist in living beings, including themselves. Because consciousness does not fit into their mechanistic schemes, they declare it illusory. Magicians make exactly the reverse argument. Observing consciousness in themselves and animals, they are magnanimous enough to extend it to all things to some degree—trees, amulets, planetary bodies, and all. This animistic or panpsychic perspective is a far more respectful and generous attitude than that of religions, most of which won't even give animals a soul.

The magical view of mind differs radically from scientific and religious ideas. From a religious point of view, we are variously the willing, unwilling, or unknowing playthings of the gods. Alternatively, we are partly of God and partly of the devil, or partly of God but mostly evil by choice. Again, moralistic thinking obscures ignorance. There is actually no scientific view of mind at all; there is only psychology, so we must contrast this with materialistic views in general.

The contrast is an odd one. Psychology claims that when something happens to people (stimulus), they do something (response). What causes one person to give a particular response and another person a different one is their ego. The general materialistic view, on the other hand, is the assumption that we have free will. Am I my ego, or am I my free will? This ancient problem is insoluble because it is wrongly phrased.

Magic offers an alternative view. Consciousness occurs when the Kia (which is equivalent to free will and perception, but is itself formless) touches materiality (the ego, mind, sensory and extrasensory information, etc.). So, we have both of these things, but we are neither of them; we experience our being only at their place of meeting.

A general overview of the magical interpretation of existence appears in the following chapters on chaos, Baphomet, the psychic censor, and Choronzon. A more detailed exposition of technical occult theories appears in the section on magical paradigms.

Chaos: The Secret of the Universe

MIGHT IT NOT BE THAT CONSCIOUSNESS, magic, and chaos are the same thing? Consciousness is able to make things happen spontaneously without prior cause. This usually happens within the brain, where that part of consciousness we designate "will" tickles the nerves to make certain thoughts and actions occur. Occasionally, consciousness is able to make things happen spontaneously outside the body when it performs magic. Any act of will is magic. Conversely, any act of conscious perception is also magic; an occurrence in nervous matter is spontaneously perceived in consciousness. Sometimes that perception can occur directly without the use of the senses, as in clairvoyance.

Magic is not just confined to consciousness. All events, including the origin of the universe, happen basically by magic. That is to say, they arise spontaneously without a final prior cause. Matter gives the appearance of being governed by physical laws, but these are only statistical approximations. It is not possible to give a final explanation of how anything happens in terms of cause and effect. At some level, the event must have "just happened." This might seem to give rise to a completely random and disordered universe. Not so. Throw a single die and you might get anything; throw six million and you will get almost exactly a million sixes. There is no reason for the laws of the universe represented here by the structure of the dice; they, too, are phenomena that have just arisen spontaneously and may one day cease to apply if spontaneity produces something different.

Now it is very difficult to imagine events arising spontaneously without prior cause, even though this happens every time one exerts one's will. For this reason, it has seemed preferable to call the root of these phenomena chaos. It is impossible for us to understand chaos, because the understanding part of ourselves is built out of matter, which mainly obeys the statistical form of causality. Indeed, all our rational thinking is structured on the hypothesis that one thing causes another. It follows then that our thinking will never be able to appreciate the nature of consciousness or the universe as a whole because these are spontaneous, magical, and chaotic by nature.

Now it would be unjustified to infer from this that the universe is conscious and can think, in our sense of the word. The universe *is* the thoughts of chaos if you like. We may be able to understand the thoughts but not the chaos from which they arise. Similarly, we may be accustomed to being conscious and exerting our will, but we shall never be able to form ideas of what these are.

Each of the major human philosophies attempts to answer a particular question about existence. Science asks "how" and discovers chains of causality. Religion asks "why" and invents theological answers. Art asks "which" and comes up with the principles of aesthetics. The question that magic seeks to answer is "what," and it is thus an examination of the nature of being or doing.

If we proceed straight to the heart of the matter and ask of magic what is the nature of consciousness and the universe and everything, we get this answer: they are spontaneous, magical, and chaotic phenomena. The force which initiates and moves the universe, and the force which lies at the center of consciousness, is whimsical and arbitrary, creating and destroying for no purpose beyond amusing itself. There is nothing spiritual or moralistic about chaos and Kia. We live in a universe where nothing is true, although some information may be useful for relative purposes. It is for us to decide what we wish to consider meaningful or good or amusing. The universe amuses itself constantly and invites us to do the same.

I personally applaud the universe for being the stupendous practical joke that it is. If there were a purpose to life, the universe, and everything, it would be far less amusing. We could only go sheepishly along with it or fight a heroic but futile battle against it. As it is, we are free to grasp whatever freedoms are available and do whatever we fancy with them. It may be that theology and even metaphysics are just bad lyric poetry, but here goes anyway:

Chaos—the word must be spoken, though only the untruth of it shall be known.
The blasphemy of it shall be our liberation.
Change is the only constant phenomenon.
Oh, let me worship the randomness of things, for all that I have ever loved has come forth from it and will be taken away by it. Chance!
Hail also unto apparent order, for it increaseth the possibilities of chaos.
There can be no absolute truth in a universe of relativities.
All things are arbitrary; some things have relative truth for a time.
Life being accidental, we are free to give it any point we like.

I do not find it necessary to account for my actions even unto myself.
I require no justification.
That I do it, is sufficient in itself.
Life is its own answer, my spirituality is the way I live it.
I will believe whatever brings me joy, power, and ecstasy.
Understanding cannot understand itself.
Perception cannot perceive itself.
Will cannot unwill itself.
The Secret of the Universe *is* the Secret of the Universe, known to me in the silences and in the storms.

Baphomet

"WHAT IS GOD?" we may well ask, since the question has been obsessing our species ever since it came up with the notion. The question has become a whole new ball game since the invention of the telescope. If the earth were reduced to the size of a grain of sand, then the universe would still be unimaginably vast on the same scale. The furthest observed objects would still be a long way off, not yards or even several miles, but still at least several thousands of billions of miles. Our world, a grain of sand in a thousand billion miles of space. It is most unlikely that whatever creates on this scale takes a personal interest in what we had for breakfast.

As a species, we started to form pretentious theories about cosmic gods only when we mixed up our own megalomaniac psychology with the vestiges of shamanistic knowledge. The monotheistic God is only an idealized image of ourselves or our fathers or our kings writ large. The perspective of the telescope now indicates that this idea was childishly small. No wonder the Inquisition burned astronomers.

Nevertheless, before the monotheistic errors were made, our species had arrived at a sophisticated appreciation of the psychic structure of our own little corner of the universe. In the first shamanistic aeon, men recognized the animating spirit of living beings. It was often depicted as the Horned God, a human with antlers. It was a force without morality, and it could not be bargained with nor placated. However, by careful observation, mediation, and training, it was possible to give oneself and one's tribe a psychic edge in a hostile environment by its appreciation.

These early psychic abilities, coupled with a high intelligence, rapidly made the puny humans into the planet's most successful species. The force that made this possible was universally symbolized as the Horned God. Horned because it conferred certain powers over animals and a horned human because it represented something extra which humans could acquire. The double horns symbolize the bipolar nature of a force which was both good and evil, light and dark, beautiful and terrible. Furthermore, the horned-god image gives an impression of the awesome and fearful nature of this type of power.

Agriculture and the beginning of settled life in city states ushered in the pagan aeon. Humanity lost touch with many aspects of this force, which related directly to nature, and began to construct all manner of improbable polytheistic and pantheistic theories to account for the behavior of itself and its environment. Knowledge became fragmented, and aspects of the force were personified as various deities. Superstition and mere religion became rife. The original magical lore and abilities survived in places but became unofficial or even went underground.

In the monotheistic aeon, religion became a fully institutionalized instrument of the state. The singular gods of this period were designed to give divine sanction to the secular and priestly powers and to provide a model for the ideal citizen. The ancient magical life force could hardly supply the basis for these new gods. Instead, Yahweh, Jehovah, Allah, and Buddha were defined as male humans idealized in terms of particular cultural ideas. Magic became a suppressed activity because the priests of the new religions were not very adept at it, and they were not prepared to risk anyone else usurping their limited abilities.

Because of an idealized conception of the unitary gods, all that was non-ideal or evil became lumped together in the form of various devil images. The Horned God of antiquity reappeared as the antigod of these systems. Its devotees met secretly as witches and sorcerers to practice their magic.

In the atheistic aeon, through which the leading earthly cultures are now passing, God became human, stripped of psychic and mystic capacities but provided with physical technology instead. By a supreme act of selective inattention, atheistic cultures manage to not observe the manifestation of any order of reality beyond the physical. The life force of the cosmos and the beings within it elude their equations and become the hidden god.

In the chaoist aeon, on whose threshold we stand, a new conception of psychic reality is forming. This new conception is growing on a number of fronts. The leading edge of quantum physics seems to be providing a theoretical basis for many of the phenomena rediscovered by the renaissance of interest in parapsychology and ancient magical practice.

In this new paradigm, the animating force of the entire vast universe can be called chaos. It is the inexpressible pregnant void from which manifest existence, order, and form arise. Being omnipresent and nondualistic, it is virtually imperceptible, inconceivable, and impossible to visualize. Almost any attempt to say anything about it would be a denial of its other qualities and so a lie.

We could say it was chaotic or random, for form arises from it without cause. We could describe it as fortuitously random, but that would only reflect our positive attitude to existence if we take the trouble to maintain one. We could say that it operates at the quantum (subatomic) level and within the core of our being, if only because we are unable to detect more than its secondary effects elsewhere. We could say that its most obvious manifestation is change. This rather effective definition is actually based on a trick or an approximation. Nothing we can know is actually static or unchanging. We would be completely unable to perceive something that was totally motionless, for it would not emit energy nor impede the flow of objects through it.

Chaos might be better visualized as the only point at rest, the "unmoving mover," as it were. However we choose to see it, the ultimate ground of being is utterly void to our understanding, impersonal and inhuman, whimsical and capricious and far too infinite and incomprehensible to be much use as a god to limited dualistic beings like ourselves.

There is a part of chaos which is of more direct relevance to the magician. This is the spirit of the life energy of our planet. All living beings have some extra quality in them which separates them from inorganic matter. The ancient shamans mainly sought to represent this force by the Horned God. In more modern times, this force has reasserted itself in our awareness under the symbol of Baphomet.

Baphomet is the psychic field generated by the totality of living beings on this planet. Since the shamanic aeon, it has been variously represented as Pan, Pangenitor, Pamphage, All-Begettor, All-Destroyer; as Shiva-Kali—creative phallus and abominable mother and destroyer; as Abraxas—polymorphic god who is both good and evil; as the animal-headed devil of sex and death; as the evil Archon set over this world; as Ishtar or Astaroth—goddess of love and war; as the *anima mundi* or world soul; or simply as "goddess." Other representations include the eagle, or Baron Samedi, or Thanateros, or Cernunnos—the horned god of the Celts.

The appellation "Baphomet" is obscure but probably arises from the Greek *Baph-metis*, union with wisdom. Gods with Baphometic names and images reoccur throughout Gnostic teachings. No image can fully represent the totality of what this force is, but it is conventionally shown as a hermaphroditic god-goddess in the form of a horned human that includes various mammalian and reptilian characteristics. It should also resume protozoan, insectivorous, and floral symbolism, for it is the animating spirit of everything from a bacterium to a tiger. If we succeed in creating machine consciousness, it will have to include mechanical elements as well.

Between its horns a torch is usually positioned, for spirit is most easily visualized as light. The image should also include necrotic elements, for it also encompasses death. Life and death are a single phenomenon through which the life force continually reincarnates. A denial of death is also a denial of life. The cellular mechanisms which allow life also make death inevitable, essential, and desirable. All religions which deny death are basically antilife. Have no fear—you have been, and will be, millions of things; all you will suffer is amnesia. The sexual aspects of the god-goddess Baphomet are always emphasized, for sex creates life, and the sexuality is a measure of the life force or vitality, no matter how it is expressed.

The spirit of the life force is the spirit of the dual ecstasy, procreation and reabsorption, sex and death. Beautiful and terrible god of the hovering hawk, god of the thrusting sapling, god of conjoined lovers, god of the worm-filled carcass, god of the starting hare, god of the wild hunt carousing the forest in mad exhilaration. Invoke this god with wild uninhibited love play, and with wine and strange drugs which thrill and exalt the vitality and imagination. Lastly, draw thine own exhilarated consciousness into communion with this god by profound concentration and visualization, and the magic life force is thine to wield for good or ill. Oh, come forth in horned majesty as the power of the air, and grant us the power of the windsight and the windspeech!

Virtually all mythologies retain some lore about primitive reptilian energies, which often antedate the gods themselves. Thus, in many cosmologies we have various Leviathan-like serpents encircling the universe, or Tiamat-type chaotic dragons from which existence springs. The gods are frequently depicted as having slain or imprisoned these reptilian forces, or as being perpetually engaged in suppressing them. Almost all demons are depicted as part animal, and the majority have some reptilian features.

A number of ingenious but incorrect suggestions have been put forward to account for the ubiquitous representation of primal or evil forces with reptilian symbols. It is true that serpents resemble the phallus, but the majority of quadrapedal reptiles do not. It is also true that some snakes give the appearance of regenerating themselves when they shed their skin, but even casual observation would quickly show that this does not make snakes immortal. Some large reptiles are undoubtedly dangerous to health, but the terrifying dinosaurs were already long extinct before man appeared.

If we do retain ancestral memories of fighting fearsome animals, then those animals would have been almost entirely mammals: mammoths, bears, aurochs, and great cats. No, there must be some more profound connection between man and the dragon to explain the universal occurrence of this cultural myth, even in lands with few spectacular reptiles.

The dragon of our mythologies sleeps inside our own heads. Evolution has left us with three brains. Instead of a complete restructuring of the brain at each phase of evolutionary advance, new bits were simply added on to cover new functions. The newest part of our brain is what makes us uniquely human. Only the apes show anything similar. The next oldest part is something we share with mammals generally. The most primitive part of the brain is something that mammals, including ourselves, share with the reptiles. The human has an ape, a wolf, and a crocodile living inside its skull.

All the dragons, serpents, and scaly demons of myth and nightmare are reptile atavisms arising out of the older parts of our brains. Evolution has not deleted these ancestral behavior patterns, merely buried them under a pile of new modifications. Thus, in mythology, the gods, as representatives of human consciousness, suppress the titans and dragons of the older consciousness.

The traditions of magic preserve a number of techniques for arousing the sleeping dragons and wolves of the older brains. If the body's aetheric forces are directed upward into the cranium, the first parts of the brain to be activated will be the reptilian circuits. Thus, in Oriental esotericism, liberating the serpent power is called raising the Kundalini. Mindful of the dangers in this technique, the Oriental magicians insisted that the Kundalini must not be allowed to linger here, but must be made to enter the higher cerebral centers.

The older brain centers can also be activated during intense states of excitement or meditative quiescence. Gnosis can be directed to these levels by visualizing oneself in the required beast-form, and by using sigils to reach the subconscious behavior programs. The "dragon mind" finds magical application in the creation of powerful and rather nasty demons and for the projection of enchantments of a similar nature. The programs of the reptilian consciousness do not extend to compassion or conscience and contain only enough forethought for the necessities of hunting, killing, eating, and reproducing.

Society and religions have been concerned with keeping the dragon and wolf permanently suppressed, except at such times as it suited them to make war. For the magician, these atavistic forces are a source of personal power. Thus Baphomet, the magicians' god, is frequently shown in composite human-mammal-serpent form, as are many shamanic gods.

The Psychic Censor

THE PHYSICAL PART OF OURSELVES is very touchy about chaos and magic; in fact, our mind abhors these things, and there is a very powerful censor mechanism that prevents us from using or noticing all but a small fraction of it.

When people are presented with real magical events, they somehow manage not to notice. If they are forced to notice something uncontrovertibly magical, they may become terrified, nauseated, and ill. The psychic censor shields us from intrusions from other realities. It edits out most telepathic communication, blinds us to prescience, and reduces our ability to register significant coincidences or recall dreams. The psychic censor is not just put there out of divine malice; ordinary physical life would be impossible without it. It would be like living permanently under the influence of hallucinogens.

The consciousness-force in us that appears as the root of will and perception can be called Kia. This Kia has no form. Any form of innate or divinely sanctioned motive that one may seem to have found in it is an illusion. It is this void at the center of one's being which is the real Holy Guardian Angel. The psychic censor, on the other hand, is a material thing which protects the mind from magic and from being overwhelmed by the awesome strangeness of the psychic dimension which appears to us as chaos.

Magicians have a number of tricks up their sleeves for selectively bypassing the psychic censor. The censor is more active on some levels of consciousness than others. On the dream level, the perception, and sometimes the will, has more freedom to act magically, but the censor will often succeed in either preventing the command to do this from penetrating the dream level or preventing the memory of it being available on the awareness level.

The awareness level—on which we are conscious of thinking and being emotional—is given the greatest degree of protection by the censor, and many magical techniques are designed to draw consciousness away from this level. The robotic level, on which we perform automatic tasks, is less well protected. In a state of absentminded preoccupation, strange flashes of almost subliminal perception may occur, but the censor often acts to prevent these from fully entering awareness. If this barrier can be overcome, an almost maddening volume of telepathy, short-term precognition, and improbable coincidence can be perceived. The gnostic level of quiescent concentration or ecstatic excitement is the least protected by the censor because in this level, much of the mind is silenced. Consequently, most effective magical systems have developed one or more methods of entering this level deliberately.

The Demon Choronzon

A CURIOUS ERROR HAS ENTERED into many systems of occult thought. This is the notion of some higher self or true will, which has been misappropriated from the monotheistic religions. There are many who like to think that they have some inner self which is somehow more real or spiritual than their ordinary or lower self. The facts do not bear this out.

There is no part of one's beliefs about oneself which cannot be modified by sufficiently powerful psychological techniques. There is nothing about oneself which cannot be taken away or changed. The proper stimuli can, if correctly applied, turn communists into fascists, saints into devils, the meek into heroes, and vice versa. There is no sovereign sanctuary within ourselves which represents our real nature. There is nobody at home in the internal fortress.

Everything we cherish as our ego, everything we believe in, is just what we have cobbled together out of the accident of our birth and subsequent experiences. With drugs, brainwashing, and other techniques of extreme persuasion, we can quite readily make a human a devotee of a different ideology, the patriot of a different country, or the follower of a different religion. Our mind is just an extension of the body, and there is no part of it which cannot be taken away or modified.

The only part of ourselves which exists above the temporary and mutable psychological structure we call the ego is the Kia. Kia is the deliberately meaningless term given to the vital spark or life force within us. The Kia is without form. It is neither this nor that. There is almost nothing we can say of it except that it is the void center of consciousness, and it "is" what it touches. It does not have any qualities like goodness, compassion, or spirituality, nor their opposites. It does, however, give a feeling of meaning or consciousness when we experience or will anything, and it becomes more apparent to us when we experience something powerfully. Laughter in ecstasy gives us a glimpse of it.

The center of consciousness is formless and without the qualities of which mind can form images. There is no one at home. Kia is anonymous. We are an incomprehensible biomystic force field, from hyperspace, if you like, with a mind and body attached. The mistake of so many occult systems is to imagine that the Kia has some preordained or intrinsic quality or nature. This is just a wishful thinking, trying to give cosmic significance to the ego. Our ego is what the mind thinks we are. It is an image of ourselves which grows out of our life experiences—our body, sex, race, religion, culture, education, socialization, fears, and desires.

There is a great pressure on us to develop an integrated and assertive ego. We are supposed to know exactly who we are and what we believe and are supposed to be able to defend that identity. The more strongly we identify with something, the more strongly must we reject its opposite. Thus, the strongest, most obsessive egos belong to the least complete beings. For these types, there is the additional problem that to exalt any principle will eventually attract its opposite. Those who exalt strength will be drawn into a position of weakness. Those who strive for good will become embroiled in evil.

Developing an ego is like building a castle against reality. It provides some defense and a sense of purpose, but the larger it is, the more it invites attack, and, ultimately, it must crumble. There is a further problem. All fortresses are also prisons. Because our beliefs imply a rejection of their opposites, they severely restrict our freedom.

Most mystics and religiously oriented magicians describe their mystical experience in terms of transcendence. They describe themselves as having been swept up into something far greater, as a leaf in a hurricane or as a teardrop slipping into an ocean. They claim that their own ego has been obliterated and merged into union with godhead.

Nothing of the sort has occurred. They have merely employed some form of gnostic exaltation to inflate their own ego into an immense version of a god that they have been carefully cultivating. The process differs not one whit from that employed by the black magician who also inflates his ego to cosmic dimensions, save that the religious types need a god in whose name to advance their own interests. They may also make a passing show of humility to conceal from themselves the enormity of their megalomania.

Exactly the same thing happens when a magician attempts to invoke a personal Holy Guardian Angel. The source of consciousness exists as the powers of will and perception. Any names, images, symbols, and directives that the magician receives will only be exaggerated artifacts from the magician's own mind and ego and possibly telepathic fragments from other people. Because these communications occurred in a gnostic state, they were likely accepted uncritically. Gnosis also unleashes subconscious creativity, and the messages are likely to be even more alluring if they are strung together with unexpected cleverness.

We, each of us, have a real Holy Guardian Angel, or Kia, which is our power of consciousness, magic, and genius. We also have a regrettable capacity to become obsessed with the mere products of our genius, mistaking them for the genius itself.

These obsessional side effects have a genetic name, Choronzon, or perhaps the demons Choronzon, for its name is multiple. To worship these creations is to imprison oneself in madness and to invoke eventual disaster.

Belief in a god or belief in one's ego are the same thing. Every human is already its own diseased vision of God. Both the religious maniac and the black magician acquire a certain charisma and mission from their respective obsessions, but ultimately their quest is futile, for they cannot get beyond their own inflated fears and desires to the real thing—the anonymous and formless, yet fantastic, power source within themselves.

That we are conscious, magical, and creative is the most mysterious and incredible thing in the universe. Any god or higher self we can imagine is necessarily less amazing than what we ourselves actually are, for it is merely one of our own creations. Myself, I am unwilling to give any sensible name, attribute, or glyph to the infinite mystery within the core of my consciousness and behind the illusion of the universe. It has been wisely said that the Absolute is either ineffable or it is less than us.

To invoke the real Holy Guardian Angel (or Kia) is a paradoxically difficult task. As it has no form, there is no way to get an imaginative grip on it. It cannot be willed or perceived, for it is, itself, the root of will and perception.

If one invokes the Holy Guardian Angel with the general expectation of various signs and manifestations, then one's genius and magical capacities will usually provide these if enough gnosis is employed. Alternatively, if one enters an exalted state in an unplanned way, then the free belief generated will usually attach itself to any incipient mystical ideas one may have. In both cases, one has missed the boat. Let me repeat my startlingly simple message. The real Holy Guardian Angel is just the force of consciousness, magic, and genius itself—nothing more. This cannot manifest in a vacuum; it is always expressed in some form, but its expressions are not the thing itself.

There are perhaps only two things one can do to invoke the real Holy Guardian Angel or Kia. Firstly, the ego can be put in its place by deliberately seeking union with anything one has rejected. Secondly, the hidden god force Kia can be felt as the root of all acts of consciousness, magic, and genius by performing as diverse and extensive a series of these acts as possible.

Invoke often, as the oracle said.

And banish Choronzon whenever it manifests.

ANIMISM AND
SHAMANISM

Animism is our oldest magical and mystical tradition. It is from animism that all religious arts and magical sciences originate. The animistic traditions are still practiced on all the southern continents—Australia, Africa, and South America. It is primarily found in hunting societies but survives also in semi-settled village life where it takes on more of the character of witch doctoring.

Shamanism is a form of animism prevalent in northern latitudes that emphasizes trance as a means of communicating with the "mana" and "spirits" of natural phenomena, plants, animals, and people. The encroachments of modern civilization have almost destroyed shamanism in North America, Northern Asia, and within the Arctic Circle. Some animistic and shamanic knowledge survived in European witchcraft, while in the Middle East, animism became swallowed up in the priestly cults of classical civilizations.

Two conclusions can be drawn from an examination of remaining animistic cultures and from records of those now extinct. Firstly, despite the enormous geographical separation between animistic cultures, they share almost identical methods. Secondly, it is animistic and shamanic knowledge and power that contemporary magicians seek to rediscover.

The basic principles of magic, like the basic principles of science, do not change, but they can become lost. Animism and shamanism present a very full magical technology which resumes all occult themes. Mankind now stands in greater need of these abilities than at any time since the first aeon if it is to understand rather than destroy itself and the environment. These traditions once guided all human societies and kept them in equilibrium with their environment for thousands of years. All occultism is an attempt to win back that awesome lost wisdom. Let us look, then, at what these traditions hold.

Shamanic power cannot be progressively accumulated like other technology. Shamans will be lucky if their apprentices make any advance beyond their own achievements. Shamanic powers are so difficult to master that a tradition requires a continual influx of talent just to prevent itself from degenerating. For this reason, shamans usually describe their tradition as having declined from past glories. Only an occasional, exceptional practitioner can win back some of the more legendary powers.

Central to shamanism is the perception of an otherworld or series of otherworlds. This type of astral or aetheric dimension, containing various powers, entities, and forces, allows real effects to be created in this world. The shaman's soul journeys through this dimension while in an ecstatic or drug-induced state of trance. The journey may be undertaken for divinatory knowledge, to cure sickness, to deliver a blow to enemies, or to find game animals.

Prospective shamans are usually selected from those with a nervous disposition. They may either be assigned to shamanic instruction or be driven to it by a power present in the shamanic culture. Initiation invokes a journey into the otherworld, a meeting with spirits, and a death-rebirth experience. In the death-rebirth experience, the candidate has a vision of his body being dismembered, often by fantastic beings or animal spirits, and then reassembled from the wreckage. The new body invariably contains an extra part, often described as an additional bone or an inclusion of magical quartz stones or sometimes an animal spirit. This experience graphically symbolizes the location of the aetheric force field within the body or the addition of various extra powers.

In most shamanic systems, this aetheric force is exuded through the navel region for short-range magics, although it can be sent through the eyes or hands instead. It is the same as the chi or ki or Kundalini or aura.

The shamanic tradition exhibits a full range of magical themes. Exorcism and curing are the main skills shared with the community, and these are usually undertaken in trance and ecstatic states during which an otherworld journey is made to seek a cure. Magical attack and protection may be performed for clients, and shamans themselves will frequently fight each other for supremacy, often assuming their otherworld animal shapes for this purpose.

Some shamanists cultivate enormous physiological control with which to resist extremes of heat, cold, and pain. Fire walking, in which fierce heat is magically prevented from scorching flesh, is a very common feature of this tradition and occurs worldwide.

Congress with the spirit world is extensive and includes various nature spirits, animal and plant entities and servitors, the shades of the dead, sexual entities like incubi and succubi, and usually a horned god, even in lands with no horned animals. Egress into the otherworld is made through perilous clashing gates, comparable to the modern conception of the Abyss. Dream as well as trance is an important method of obtaining access to the otherworld.

Shamanic tools are highly varied but usually include a noisemaking device, such as a drum or snake-bone rattle, to call spirits and induce trance, as well as various power objects, most commonly quartz crystals. The extraordinary traditions of shamanism are the fountainhead of all occult systems, and it is to animism and shamanism that we must look if we wish to pick up the pieces of magic, man's oldest science, and use them again.

GNOSTICISM

In the 1st and 2nd centuries, a series of bizarre cults sprang up in various parts of the Roman Empire, notably in Alexandria, that melting pot of peoples and cultures at the mouth of the Nile. These cults were known as Gnostic. Their ideas and activities seem oddly to be both antique and highly advanced.

When the black order of hierarchical Christianity rose to ascendency, it vigorously and violently suppressed these cults. However, you cannot blame Jesus for the religion practiced in his name. The Gnostics left a wealth of written material, and some of their cults survived underground to influence the development of the magic art in later centuries. The medieval Cathars and Albigensians certainly possessed some Gnostic knowledge, and this chapter will suggest that their influence can be detected at many other points.

There are many threads in Gnostic thought. It contains cosmic speculations elevated enough to compare with the most refined of the Eastern systems. Some of these speculations anticipate medieval Kabbala and astrology. There is a well-developed system of magic surviving mostly in the form of artifacts. The Gnostics had a variety of ethical systems based either on complete anarchic libertinism or else on strict asceticism, whichever seemed most likely to lead to liberation in any particular situation. Above all, Gnosticism was concerned with mystical experience—gnosis—as opposed to mere *pistis*, or faith. What the world has tended to remember the Gnostics for, however, is their apocryphal stories, which mock the orthodox religions of their times.

Gnosticism has been supremely important in the development of Western occultism, for it represents a synthesis of Greek, Egyptian, and Oriental enlightenments, which were swiftly forced underground and later appeared in the works of medieval and renaissance magicians, in the Templars, in witchcraft, in Rosicrucianism, and in our own time.

To the Gnostics, no conception of God or the ultimate or whatever was infinite enough. They considered that the Supreme Being was completely ineffable and beyond anything that could be said of it. They laughed at the hopelessly parochial anthropomorphic conceptions of the Absolute that other religions put forward and endeavored to say as little about it as possible, save that it was too immense to have ideas about. To them it was like the Tao or the void. They did, however, consider that there was a small fragment of this infinitude in humans and in every living being. Gnosis meant experiencing this primal spark within oneself.

Exactly how the infinite fragmented itself and descended into existence with matter was the subject of unending debate amongst the Gnostics. They produced many theories. Some were merely poetic allegories of the process in sexual terms. Some were allegorical commentaries on human psychology—every cosmology embodies a psychology. Some were excuses to heap ridicule on other religions. Some were probably deliberate attempts to ridicule the idea of understanding the process with the mind at all. In constructing these theories, they produced a varied and colorful intermediate magical world of various Aeons and archons between this world and the ultimate reality.

The ultimate reality gave rise to a number of Aeons, usually thirty, which surround the material universe. These Aeons are not so much periods of time as spiritual principles or principalities. This idea seems to have reappeared in the magical visions of Dr. John Dee, who saw them as thirty Aethers. Various tensions inherent in the Aeons resulted in the formation of a number of archons, or rulers. In other systems, the ultimate reality itself is the first archon and from this a number of subsequent archons, usually seven, the Hebdomad, evolved by a process of *ennoia* or what we might call thought projection. The ennoia of the first archon produced a being, Barbelo (or Barbelon), having a female or androgynous nature. Alternatively, Barbelo might be identified with the "great silence" in which the "prime cause" or first archon manifested.

Somehow, from these cosmic principles, the force responsible for the creation of this world arose. This is variously called Ialdaboath or Sabaoth or Iao and many other names. Sometimes the force is sevenfold and identified with the astrological planets. This force is conceived of as androgynous or male with an animal-headed manifestation. It is held responsible for the creation of material beings into which the ultimate reality then condescended to breathe a vital spark.

Barbelo is known to us as Babalon or Nuit, the great star mother in whom one must seek reabsorption to penetrate the highest mystery. Ialdaboath was yet another manifestation of the ubiquitous horned god known to the Templars as Baphomet and the Christians as the devil. The dragon force appears in some gnostic systems as the world-serpent or Leviathan, encircling the universe and biting its own tail.

The Gnostics' attitudes to material life—although apparently contradictory—are a direct consequence of their gnosis and their cosmological speculations. Having experienced the spark of the infinite within, they realized they could not be touched by anything, and thus they were free to do anything at all. Some considered particular forms of activity more likely to obscure the vital spark and other forms more likely to liberate it. Some were libertines, some ascetics—they usually chose to be the opposite of prevailing social customs.

The material world was considered to be entirely evil, corrupt, and imperfect. This was chiefly because of its obvious impermanence. Only the vital spark was immortal and would reincarnate until it achieved union with the infinite, either at the end of the universe or by liberating itself in the meantime. This then, briefly, was the gnostic view of reality. Gnosticism was never an organized religion but existed as a series of elitist cults led by such notables as the wizard Simon Magus, the philosopher Valentinius, and Apollonius of Tyana.

Each teacher spread gnosis by word of mouth, couching the message in a form suited to the local belief structure, adapting gnostic practices to local need. In addition, a great deal was written down, partly to remind particular teachers what they had taught and also to sow confusion and dissent in the ranks of the main organized religions of the day—Hellenic paganism, Christianity, and Judaism.

A number of alternative bits of the Bible were produced to put across some important gnostic speculations. Firstly, the God Yahweh of the Old Testament was seen as a vicious, senile old fool intent on persecuting humanity, while the serpent (who gave knowledge) was seen as humanity's friend. Secondly, Jesus was seen as a true messenger of the infinite, but his crucifixion was considered meaningless. Only his message of love and the power above were thought important.

The Gnostics were true anarchists of the spirit. They saw all other religions as encouraging enslavement to priesthoods and secular powers with their legal and moral strictures. Against these things they pitted their cosmological jokes, their antimorality, and their magic.

Gnostic magic included the use of familiar spirits, necromancy, and the use of potions for erotic and dream-inducing purposes, but their main practices were orgiastic, telesmatic, and incantatory. Their orgiastic rites included the consumption (as sacraments) of the mixed male and female sexual elixirs and menstrual blood after coitus. They were also reputed (probably falsely) to have consumed their own deliberately aborted fetuses. Most gnostic sects were not interested in reproduction—which they considered to be a repetition of a fundamental error. Their sexual rites were designed to cheat the evil archon of further human victims and to give an inspirational foretaste of the final and ultimate reabsorption into Babylon.

The Gnostics left behind them innumerable intricate and beautiful impressions on stone, jewels, ceramic, and metal, which go under the name "gnostic gems." These would have functioned as talismans and amulets charged with various spells and enchantments. They have also left us some very striking and bizarre votive statuary, which would have functioned as fetishistic centerpieces in rituals.

Many of the words of power and barbarous names of evocation that exist in medieval and contemporary magic have their origin in gnostic incantations. These are often woven into invocations of great beauty and power, such as the "bornless" or "headless" ritual. The word "abracadabra" comes from the name of the gnostic god Abraxas. A number of gnostic sects were active around the Damascus area, and if one were trying to rediscover or even invent the dread Necronomicon of the Lovecraft mythos, then Gnosticism would be a good source.

The timeless themes of magic appear in their completeness in Gnosticism because it was able to draw its techniques from Egyptian learning, from the Greek Mystery Schools, and from systems further east, each of which had preserved traditions from that ultimate wellspring of magic—animism and shamanism.

OCCULT
PRIESTCRAFT

M agical, mystical, and religious enterprises seek to fulfill five basic human needs, which can be identified as follows:

To provide techniques of emotional engineering.

To give life a sense of meaning.

To provide some means of intercession or intervention.

To supply an explanation of death.

To formulate a social structure or cult.

These needs are deeply interrelated, and many religions, and particularly many political philosophies, do not attempt to deal with them all. Finding a solution to some of the problems may make it less urgent to solve others. Occult priests should be capable of dealing with all of these issues. Let us consider how they might tackle each one and contrast their methods with those of the more orthodox systems.

Emotional Engineering

This includes all practices designed to stimulate or control emotional states: exaltation in prayer and song, contrition and guilt for imagined sins, fear and anguish at the specter of divine anger, and joy at the prospect of divine reward.

In our culture, a correlation between a fall-off in religion and an increase in the use of mood-altering drugs is very marked. The greatest threat to religion, however, is entertainment. The new power of the entertainment media to supply us with everything from joy to terror has usurped many of the functions of the priest. There is a refreshing honesty about secular entertainment; it's just entertainment without the excuse of spirituality for its justification. It is still, however, manipulative.

If anyone wishes to put themself into or out of any emotional state, then they should be provided with the techniques to accomplish this. The process requires no justification—that anyone wills it is sufficient. One cannot escape emotional experience in a human incarnation, and it is preferable to adopt a master rather than a slave relationship to it. The occult priest should be capable of instructing anyone in the procedures of emotional engineering. The main methods are the gnostic ones of casting oneself into a frenzied ecstasy, stilling the mind to a point of absolute quiescence, and evoking the laughter of the gods by combining laughter with the contemplation of paradox.

Anyone who masters these techniques fully has achieved a tremendous power over themself more valuable than health, love, fame, or riches. They have set themselves free from the effects of the world; nothing can touch them unless they will it. As it has been said, the sage who knows how can live comfortably in hell.

Meaning

Meaning is motivation. Anything which gives rise to physical and mental behavior of any kind is providing meaning. Thus, the body is the source of many basic meanings in this world. Pain, pleasure, hunger, sexuality, and so on provide an impetus to action and hence a source of meaning. Once the organism has solved these problems, other more subtle motivations arise on the mental level—desire for knowledge and power and emotional gratification of all kinds. Beyond this, the organism may seek higher level motivations, which have been called "spiritual," and there are some who seek the meaning of meaning itself.

To question any level of meaning with reason is usually to lose it. Meaning arises from the differentiation of experience into pain and pleasure, good and evil, interesting and uninteresting, beautiful and ugly, worthwhile and not worthwhile. Experiences are only meaningful when we are sensitive to them. We can only perceive difference. Ideas are only meaningful when we can appreciate their separateness and novelty. Spirituality only arises when we begin to consider some things nonspiritual. Meaning is dependent on establishing dualities, and belief is fundamentally an act of differentiation—considering one thing different from another.

So, ideas which create meaning for us must be conditional beliefs. For example, certain knowledge about God, whether it is yes or no, or certainty about everlasting life, either yes or no, would completely destroy any meaning in these ideas. If this were absolutely believed to be everlasting heaven or hell with no escape, there would be no reason to care about anything.

Reason is therefore destructive of meaning when it seeks unconditional and absolute answers. In this context it is probably more prudent to stay the hand of suicide and ask if reason is not somewhat out of phase with the nature of existence.

The ascetic mystic and the magician adopt different stances toward their respective existences. The ascetic mystic conceives a vast differentiation between the material and the spiritual. Mystics attempt to withdraw meaning from the material so that they can put it into the spiritual. Withdrawing meaning from the material seems a bizarre exercise, but there is an inner logic to it. Indifference to sex, indifference to hunger, to pleasure, and to pain, indeed to everything that motivates normal people, opens a whole world of "spiritual" experiences; dreams, acts of devotion, and inner thoughts become charged with meaning.

For those who devise or believe in religions, it is necessary to erect conceptions on a cosmic scale to provide a source of reference and meaning. Invariably, the highest principle must be paradoxical or contain some duality. Either the ultimate principle must actually consist of two opposed principles, or there must be some sort of fall from the ultimate. The paradoxes of religion are unquestionable and can be interpreted only on a hierarchical basis. Religions are innately repressive and conservative. Only heresy and schism permit any evolution of ideas. Much of the meaning in religion derives from authority-and-obedience relationships; hence religions exist only as social phenomena. Private religion inevitably evolves into mysticism or magic, and these have a tendency to devolve into new religions.

The magician does not conceive of such a vast gulf between spirit and matter; they are both part of the same thing with neither exalted above the other. The magician rejects no part of the experience. The magician lives in a continuum, beginning with the sublime and ineffable Tao/

God/chaos through the mysterious and subtle aethers to the awesome and strange material world. To the magician, any piece of knowledge, any new power, any opportunity for enlightenment is worth having for its own sake.

The only thing abhorrent in this incredible existence is a failure to come to grips with some part of it. To be able to operate in all spheres, the magician must master the art of either acting without belief or being able to temporarily invest belief in anything. The magician should be equally at home with a crozier, a paintbrush, a test tube, or a wand. In all things, the magician seeks to bring Kia into manifestation; for life is its own answer, and the way it is lived is spirituality.

It is senseless to ask grand and unspecific questions about life and the universe in general, because for an answer we can only invent hypothetical states of not-life or not-universe. The universe as it exists is a fantastic and magical place, and life is a mystery whose depth can never be exhausted. It is only when humans do not pay enough attention to the totality of what is going on around them every second that they are tempted to invent spurious theories to cover this lack of knowledge. For the magician, that lack of knowledge is the ultimate source of meaning. The true priest is one who can communicate this sense of mystery.

Intercession or Intervention

All religions have some method of affecting reality, or of encouraging some god to affect reality, or of merely giving the appearance that they are doing these things. As a religion becomes more institutionalized and orthodox, there is less and less emphasis on this type of activity—and for good reason. Magic is a very anarchic business.

Some people have a greater gift for it than others and sometimes it fails. Most priests who become adept at it would soon realize it was their own psychic power at work and not that of a god. Such priests who became adept would soon attract a huge following and usurp and disorder the clerical hierarchy. All orthodoxies tend to frown on the use of magic for this reason, and also because they might not themselves be able to deliver on demand.

The religions' answer is to involve the congregation in a half-hearted attempt at intercession, and then be ready with the catchphrase, "It was not God's will," in the event of failure. One might well ask, if God is going to do his will anyway, surely he does not require a cue from us.

The approach of the occult priest is entirely different when he leads his order or coven in magical activity. There is a high probability of failure because they may not be able to raise sufficient power and they may not be doing exactly what is required. Everyone will be aware of this. In this situation, they have to act with total commitment and without the slightest trace of lust of result. Everything possible must be done on the physical plane to set up the conditions of success beforehand, and then the magic is hurled in to tip the balance. To have given one's utmost is enough in itself. The result can be awaited without fear or desire and received with laughter, whatever it may be.

Death

The difference between ideas and beliefs is that ideas may be true, but beliefs are always false. That may seem a monstrous thing to say, but I offer it as a definition. What separates an idea from a belief is the emotional force committed to upholding the belief. If something were really true for us, we would not have to make an effort to believe it.

All beliefs about death have one other characteristic apart from their inherent improbability and falsity. They have to be conditional. That is to say, they have to contain both heaven and hell, or both pleasant and unpleasant reincarnations. Consider a scheme in which one was destined for either everlasting heaven or everlasting hell or for either total extinction or perpetual existence as a totally disembodied spirit without the organs of will or sense. Or consider a certain knowledge that one's next incarnation could not be affected by events in this life.

As beliefs, these things would be quite useless and unsatisfying. This reveals beliefs about death for what they mostly are, devices to create emotional effects in this life. The occult priest should abstain from adding anything to this necrotic heap. Rather, he should devote his talents to showing people what death is like.

Necromancy is something of a dying art these days, largely because it has been widely abused by those who are merely telepathic with the living and/or want money. Nevertheless, those who have directly seen or spoken to the dead have a certainty of something beyond faith. The single experience of having left one's body for a time is worth more than any belief, and it is the only useful preparation for death. The experience is reasonably accessible to any determined person.

Social Structure or Cult

Any human enterprise involving more than one individual will exhibit some form of social structure from complete hierarchy to apparent democracy. The dynamics of assorted cults, cabals, and religions are instructive of the various ways in which magical orders ought, and ought not, to be organized.

In a religion, hierarchy is of paramount importance and is effectively an object of worship itself, though this would never be openly stated. To enslave their followers, hierarchs represent themselves as the emissaries of higher powers or "the teaching" or whatever, but not as the thing itself. This is analogous to troops saluting not the officers but the commissions they wear on their chests. The end result is the same, but it does help to overcome the ego resistance involved in one person submitting to the will of another.

Once such an asymmetric relationship has been established, it readily perpetuates itself. The priest or leader is permitted to pass personal comments on followers. These do not even need to be especially perceptive. They need only be the sort of things one's friends wouldn't say to one's face, plus a few things one would like to hear, and suddenly the guru appears to be the wisest person on earth.

Another gambit of religious and political organizations is to force a re-rationalization of beliefs through action. People are not persuaded into belief intellectually. They are persuaded to perform religious acts in childhood or while under stress. Afterward, they develop or accept the rationalizations and opinions that go with it. To convert someone to anarchism, persuade them to throw a bomb for various romantic emotional reasons. They will subsequently have to adjust their beliefs to justify what they have done. The most successful organizations are those that plunge potential converts straight into action.

Obedience follows a similar pattern. At first, only the smallest and most inconsequential obediences will be demanded. These force the rationalization that one is in fact loyal to whatever one is giving one's obedience to. This loyalty is but a stepping stone to greater acts of submission, usually of one's intelligence, wallet, and sexual favors.

Leader-follower relationships also allow the leader to license followers to act without responsibility. The natural inhibitions to displays of violence, sexuality, and other emotionalities can easily be overridden if the leader tells followers to do these things. They will often give thanks for the permission to do what they have always had a desire to do.

Secrecy and elitism characterize all hierarchies. There is nothing wrong with being elite or having real secrets, as such, but most cults rely heavily on artificial elites and manufacture secrecy as a means of enticement and control. Accepting that elites exist and keeping their secrets are acts of obedience themselves. Membership within an elite and a degree of megalomania are among the licenses leaders can confer upon followers. To this end, most cults reinforce their collective identity with standards of dress and behavior and all manner of badges, insignia, and labels. These often assume as much importance as the actual activities of the cult. People are easily tricked into accepting membership in a large group as a substitute for enlarging themselves.

The activities of cults would seem to presuppose a high degree of cynicism among their leaders. This is rarely so. Most have swallowed their own lies and deceptions totally, or else rationalized them in terms of an even higher cause. As a result, their burning obsession equips them with a certain charisma which puts fire in their eyes and inflames their speech. And what is the end result of all this cultish activity?

COMMERCIALISM OR A POLICE RAID

A cult either manages to turn itself into a harmless institution, or becomes progressively more extremist until the state decides to smash it. A genuine magical order will be engaged in psychic guerrilla warfare against all black cults and religions, including materialist philosophies. In such a cult, every person is their own priest. Any member has the right to teach any other member any knowledge that they possess. No member has the right to claim any secret within the order beyond the secret of their own identity if so wished.

Unlike cults and religions, a genuine order will not admit people on a number-for-its-own-sake basis. Vitality can only be maintained by quality control at the intake. Whatever hierarchy arises within an order will be a reflection of demonstrable ability. Attempts to use the various tricks of the teacher enumerated in this section will be immediately spotted and ridiculed. There is only one justification for the existence of a genuine magical order—to enable individuals to take control of their own spirituality. And that is a very heroic and dangerous objective. Watch out for the police raid.

MAGICAL WEAPONS

The five classes of magical weapons are divided according to their function, rather than by the gross appearance they may manifest on the physical plane. All weapons are designed to have an effect on the physical, but the weapons themselves exist primarily on the aetheric or astral level. The physical form of a magical weapon is no more than a convenient handle or anchor for its aetheric form.

The sword and pentacle are weapons of analysis and synthesis, respectively. Upon the pentacle, aetheric forms, images, and powers are assembled when the magical will and perception vitalize the imagination. The magician may create hundreds of pentacles in the course of various sorceries, yet there is a virtue in having a general-purpose weapon of this class, for its power increases with use, and it can be employed as an altar for the consecration of lesser pentacles. For many operations of an evocatory type, the pentacle is placed on the cup, and the conjuration performed with the wand.

The sword, or more usually the dagger, is the weapon of analysis or scission or, in the most simple sense, destruction. Through the sword, the magical will and perception vitalize the imagination of the undoing of things. The sword is the reservoir of the power which disintegrates aetheric influences through which the material plane is affected. Both the sword and pentacle are aetheric weapons through which the higher order powers of will, perception, and imagination execute mental commands on the planes of middle nature.

The wand and cup weapons are used to transmit the power of the life force (or Kia) directly onto the aetheric. The wand is the weapon of will and the cup that of perception. These words are used to convey the indescribable processes which occur at the interface of consciousness and matter, rather than mere sense perception and motor action. All that can be said of these processes is that some events have the appearance of proceeding from the outside into us, and others appear to originate within us and proceed outward. The lesson of all higher ecstasies is that this difference is arbitrary and unreal. Here, we are entering a realm into which our logic structures are ill equipped to follow, and only the powers of the lamp transcend the paradox.

The cup can be regarded as an aetheric receptacle for magical perception. Of all the weapons, it is the one least likely to resemble the physical object whose name it bears, although actual cups of ink or blood are sometimes used. For some, the cup exists as a mirror, a shew stone, a state of trance, a tarot pack, a mandala, a state of dreaming, or a feeling that just comes to them. These things often act as devices for preoccupying oneself with something else so that magical perceptions can surface unhindered by discursive thought and imagination. Part of the power that is built up in them can be likened to self-fascination. The cup weapon acquires an auto-hypnotic quality and provides a doorway through which the perception has access to other realms.

The wand weapon similarly appears in a profusion of forms. As an instrument to assist the projection of the magical will onto the aetheric and material planes, it could be a general-purpose sigil, an amulet, a ring, an enchanting mantra, or even an act or gesture one performs. As with the pentacle, there is a virtue in having a small, portable, and permanent device of this class, for power accrues in it with use. As with the cup, the power of the wand is partly to fascinate the surface functions of the mind and channel the forces concealed in the depths. Like the sword, the wand is manipulated in such a way as to describe vividly to the will and subconscious what is required of them.

The lamp weapon is only named as such because of the popular analogy of spirit with light. Chaos, the ultimate substrate of existence, and Kia, the personal life force, are equally likely to be felt as an awesome darkness or as both brilliance and voidness simultaneously. As a device to channel these forces to the mundane consciousness of the magician, there is no limit to the forms the lamp might take. It could be anything from an idea of God or the Tao, to some primitive-looking fetish or symbol.

The way of the magician is the manifestation of spirit within matter, and his primary technique is gnosis, the focusing of consciousness by physiological means. The magician's lamp should be something which aids gnosis and receives the forces generated. The lamp is the weapon of inspiration in the original sense of the word—it inspirits the magician.

The magician should be capable of performing any ritual on the astral; that is to say, by the power of imagination alone. By strongly visualizing any of the weapons to the point of hallucination, the magician draws both the aetheric form of the weapon and the associated internal powers into action. Such empty-hand techniques are the mark of an adept.

MAGICAL PARADIGMS

Every system of thought and understanding stems from a number of basic postulates about the universe and humanity's relationship to it. These ideas and assumptions go to make up the paradigm or dominant worldview through which a culture or an individual interacts with its universe.

Aeons are marked by the passage of various great metaphysical thought paradigms, rather than by the passage of set periods of historical time. Within each great paradigm, there will be lesser paradigms which contribute to the whole. For example, in the dominant white, Anglo-Saxon, Protestant culture of Europe and America, the main paradigms are Protestant atheism, with its dependent paradigms of liberal humanistic individualism and the work ethic, and science, with its dependent paradigms of causality and materialism.

Other cultures have had, and still do have, completely different worldviews that are difficult for an outsider to enter. The universe (being the accommodating creature that she is) will tend to provide confirmation of any paradigm one chooses to live in. We are, to an extent, in an observer-created universe. Rather than just drifting into the magical worldview in a haphazard way, it is useful to consider the alternative paradigms in which we might intend to operate. As most of us already have our being within a scientific cultural paradigm, a modern magical view must also encompass this if it is to be effective in a technological civilization.

Six alternative magical paradigms appear below, and they are indeed a strange mixture of sorcery and hyperscience. All of them seem a little crazy from our normal point of view, but our normal point of view also proves to be pretty strange on close inspection.

All magical paradigms partake of some form of action at a distance, be it distance in space or time or both. Although we are unable to imagine how this can occur, we should not throw it out the window. Science can readily demonstrate both action at a distance with gravity and magnetism, and the distortion of space/time in so-called ordinary reality. In magic, this is not called ordinary reality. In magic this is called synchronicity. A mental event, perception, or an act of will occurs at the same time (synchronously) as an event in the material world. Science does not deny the possibility that pure information can be transmitted from place to place; indeed, the quantum inseparability principle demands that it must be. Of course, this can always be excused as coincidence, but most magicians would be quite content with being able to arrange coincidences. The following six paradigms seek to explain the mechanisms at work.

The Chaoetheric Paradigm

The manifest universe is just a tiny island of comparative order, set in an infinite ocean of primal chaos or potential. Moreover, that limitless chaos pervades every interstice of our island of order. This island of order was randomly spewed up out of chaos and will eventually be redissolved into it. Although this universe is a highly unlikely event, it was bound to occur eventually.

We ourselves are the most highly ordered structure known on that island, yet in the very center of our being is a spark of that same chaos which gives rise to the illusion of this universe. It is this spark of chaos that animates us and allows us to work magic. We cannot perceive chaos directly, for it simultaneously contains the opposite to anything we might think it is. We can, however, occasionally glimpse and make use of partially formed matter, which has only a probabilistic and indeterminate existence. This stuff we call the aethers.

If it makes us feel any better, we can call this chaos, the Tao, or God, and imagine it to be benevolent and human-hearted. There are two schools of thought in magic. One considers the formative agent of the universe to be random and chaotic, and the other considers that it is a force of spiritual consciousness. As they have only themselves on which to base their speculations, they are basically saying that their own natures are either random and chaotic or spiritually conscious. Myself, I am inclined to the view that my spiritual consciousness is random and chaotic in an agreeable sort of way.

Probability Manipulation

This is a far more modest, far less cosmically pretentious version of the first paradigm. There is a point somewhere in the genesis of any event when its future reality is uncertain. The universe is not an automatic clockwork structure; there is a level of disorganization within which the universe itself does not know what it is about to do. The information is not predictable, even in the events themselves.

Bizarre as it may seem, there is even an accurate mathematical formulation of the limits of this disorganization, the Heisenberg uncertainty principle. Events in the mind certainly partake of a similar quality: they are unpredictable and arise seemingly without cause. It has been suspected, even by scientists, that there is a hidden variable which causes the event to materialize in one particular form from a number of possibilities. This hidden variable is suspected of being consciousness or information. Consciousness, then, could be controlling how the indeterminate events will actually materialize. Armed with this idea and applying magic to the critical point, the magician may engineer some impressive coincidences.

Morphic Field Theory

The novel and extraordinary hypothesis of formative causation provides an excellent magical paradigm. Briefly, it states that whenever a new event occurs in the universe, it predisposes all subsequent similar events to occur in the same way ubiquitously across space and time by the agency of a "morphic field." The hypothesis does not concern itself with why the event occurred in the first place but suggests that as soon as it has happened, it generates this morphic field which makes it more likely to occur again. This provides a rationale for much magic.

Clairvoyance, for example, is the tapping of a morphic field left by an event in the distant or recent past. Only prophecy, always the most doubt-ful of the mantic arts, cannot be fitted into this scheme. Atavisms, entities, gods, and demons would represent the morphic fields left by human thoughts and animals. Sympathetic magic becomes the deliberate representation of an event in miniature to produce a morphic field, which causes the desired event to occur elsewhere. If imagining an event is enough to generate a small morphic field, then the effectiveness of visualization is explained.

Religion takes the view that consciousness preceded organic life. Supposedly, there were gods, angelic forces, titans, and demons setting the scene before material life developed. Science takes the opposite view and considers that much organic evolution occurred before the phenomenon of consciousness appeared. Magic, which has given more attention to the quality of consciousness itself, takes an alternative view and concludes that organic and psychic forms evolve synchronously. As organic development occurs, a psychic field is generated which feeds back into the organic forms. Thus, each species of living being has its own type of psychic form or magical essence.

These egregores may occasionally be felt as a presence or even glimpsed in the form of the species they watch over. Those who have perceived the human egregore usually describe it as God. Communion with the morphic fields of beasts is of great importance to the shaman and sorcerer, as it conveys intimate knowledge about the actual creature and allows the magician a certain power over the species. It may also allow the appropriation of certain of the beast's powers, particularly on the aetheric plane. This is the reason for the worldwide occurrence of totemism among hunting peoples and the prevalence of animal-headed/human-bodied gods in most mythologies.

Magicians consider that all life on this world contributes to, and depends on, a vast composite egregore, which has variously been known as the Great Mother, the anima mundi, the Great Archon, the devil, Pan, and Baphomet.

Observer-Created Universe

We usually tend to regard will and perception as separate functions of our consciousness or awareness. Indeed, our will and perception seem to be the most basic properties of our being. However, try making these assumptions:

Everything we perceive is real. (not unreasonable)

Anything we cannot perceive does not exist. (not for us, anyway)

Anything we will that does not come into our perception was not will but merely a failed wish.

Then *will and perception are the same thing.*

Glance around you for a moment; your entire universe is exactly as you willed and perceived it. It is all a creation of your belief in it. Even other persons must be counted as figments of your belief in them. Obviously, the beliefs which uphold the universe must be pretty deep-seated and not amenable to mere wishes, although real acts of will/perception might change parts of it. This provides a magical model in which everything is permitted, though it might be damnably difficult.

Austin Spare often worked within this paradigm, anticipating by half a century the development of nonobjectivity, one of a number of interpretations of quantum theory. This suggests that it is the act of willed perception or measurement which actually creates events. Magically, it is by tapping the deepest levels of consciousness and belief that creative events are initiated.

The Holographic Universe

To specify the position of a particle with complete accuracy, we would have to also specify the relative position of each and every other particle in the universe. In the holographic universe model, this idea is taken one step further; every particle in the universe is actually connected to every other particle by some hidden form of instantaneous connection. This connection has a mathematical formulation in the quantum inseparability principle.

Hypotheses of this type are often called "bootstrap theories" because they suggest that everything is the cause of everything—the universe is holding itself up by its own bootstraps. Any change anywhere in such a holographic universe would, in theory, be detectable everywhere instantly. Such a hidden form of instantaneous communication is the very stuff of magic.

The web of connections between every event can be seen as a higher order reality, the hologram. The part of reality of which we are normally aware is merely a projection of this, the holograph. Synchronicity and all other magical paradigms assume that there is some form of information transfer which can proceed in rather unusual ways across space and time.

Although it is difficult to imagine how matter or energy can behave in this way, there is no reason why pure information itself cannot be made to do this. Pure information has no weight or force, so nothing could prevent its instantaneous passage to anywhere or perhaps "anywhen." It is likely that somewhere within the psyche and within the quantum indeterminacy that underlies physical reality there may be something acting as transmitter and receiver of this pure information.

This would, for example, explain why psychic phenomena can be perceived but cannot easily be made to register objectively, as anyone who has tried to photograph or tape record ghosts will appreciate. It would also confirm the magical commonplace that it is easier to make a man fall under a sixteen-ton weight than to make a sixteen-ton weight fall upon a man. The information required is infinitely smaller in the first case, unless of course one can hex the crane driver at exactly the right moment.

Higher Dimensionality

We find ourselves in a universe that is at least four-dimensional. To be sensible to us, an event must have a displacement in both space and time. A piece of paper having only two or three dimensions, that is having no thickness or existing for an imperceptibly short time, could not be part of our universe. Although we commonly think in terms of a three-dimensional reality, this must be at least a four-dimensional reality, even if time does appear to have a different quality to our perception. We often forget to include time in our conceptions because we take simultaneity for granted; we assume that things exist in the same time frame and that they will persist.

Suppose for a moment that there were a fifth dimension that we were unable to perceive with our senses. Suppose also that all phenomena were five-dimensional. How could five-dimensional phenomena appear to our four-dimensional perception? A point from the fifth dimension would not be perceivable in four dimensions. This may explain why fundamental particles or quarks cannot be observed singly.

Higher dimensionality might also solve other problems in physics, like parity violation and certain properties of the vacuum. A line from the fifth dimension would appear to us only as a point. A five-dimensional surface would appear as a line in our reality. A five-dimensional "solid object" would be perceived as a surface, and the things which appear to us as solid objects in our reality would be the manifestations of true five-dimensional solids for which we have no name. Now, points, lines, surfaces, and "solid objects" do not actually exist in this fifth dimension any more than points, lines, and surfaces exist in ordinary four-dimensional geometry, except as idealizations.

It therefore follows that if there were five dimensions, all the things we perceive as existing solid objects in "our reality" are merely the way that five-dimensional objects appear to us. This may seem a nonsensical complication, but it was raised to demonstrate that we might be living in a five-dimensional reality, yet be unable to perceive it as such. What would be the consequences if this were so? It would actually explain a great deal beyond some obscure problems in fundamental physics.

Principally, it would explain why we seem to live in a world of effect rather than a world of cause. We seem only able to measure effects. We have no idea how anything causes anything else in a final sense. All our so-called physical laws are merely catalogs of effects we have come to expect. Our power to actually cause events is illusory. We merely arrange things to make certain effects more probable, but we can't get hold of the root causes themselves. This is hardly surprising if we are unable to interact with the full dimensionality of an event. As the Kabbalists have said, the causal world exists as a hidden dimension.

A fifth dimension to which the psyche had some limited access could explain all magical and occult phenomena without exception. Information moved through a fifth dimension could manifest at any point in ordinary time or space. Telepathy, necromancy, clairvoyance, and precognition are explained at a stroke. Transformations caused in the fifth dimension would appear as effects in ordinary reality; telekinesis and all forms of spell casting and enchantment are thus possible. Trying to make things happen in the ordinary world by arranging for effects is a laborious and time-consuming business. If we could gain access to the causal world, infinite power and possibility might be available at a whim, if we were still interested.

The purpose of this study is not to rehabilitate science and magic, but to demonstrate that there are alternatives to irrationalism when it comes to erecting the theoretical basis of a magical *modus operandi*. If science ever did begin to make serious enquiry into magic, the result would be disaster. Humanity has proved itself totally incapable of handling even a moderately dangerous substance like plutonium with responsibility. Imagine what it would do with machine-enhanced sorcery or even simple, reliable telepathy. It is in the interests of the survival of the species that occultists continue to ridicule and discredit their own arts in the eyes of orthodox science.

The author has a certain preference for paradigms of higher dimensionality, if only because the evolution of the simplest regular forms through increasing dimensionality leads one to profoundly familiar symbols shown together in figure 1 on page 179.

Most magical paradigms envisage a total universe made up of three realities.

Primary Reality: the void, chaos, Ain Soph Aor, God, the Empyrean, universe B, the Mean, the Pleroma or Plenum, Mummu, the Nagual, the archetypal or formative world, the fifth dimension, cosmic mind, the hologram, the night of Pan, hyperspace, acausality, quantum realm.

Secondary Reality: The aethers or astral, probability, the gods, morphic fields, the shadow world, the side, the wind, the astral light, potentia, aura, middle nature.

Tertiary Reality: The physical or material world, Malkuth, universe A, the tonal, the fourth dimension, the body of God, the holograph, causality.

And it is a further characteristic of all magical paradigms that there is an equivalence between the microcosm and macrocosm. As above, so below. Thus, humans contain a part of primary and secondary reality in addition to their physical being.

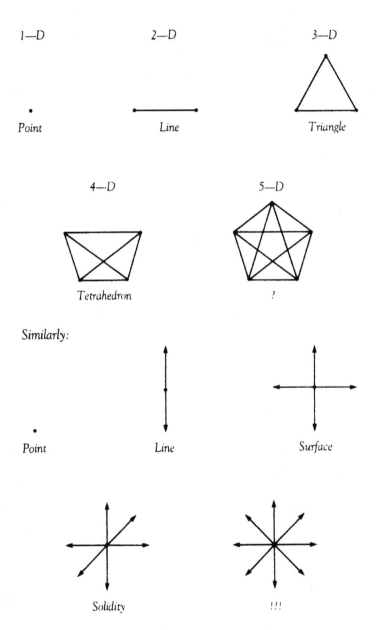

Figure 1. *The evolution of forms through various dimensional levels.*

ANECDOTES

Each of these stories from the author's casebook is interesting, because it illustrates either a particular technique or the genesis of particular ideas appearing in this book. In some cases, details of precise locations are omitted or names changed to protect individuals. As far as possible, events are presented exactly as they happened to me.

Perhaps it was because I was born and bred near to that part of the South Downs that holds both the remains of an ancient Romano-British temple at Chanctonbury and a Templar stronghold at Steyning. Or perhaps it was because I have a psychic mother. I don't know, but I can still remember meeting strange entities in childhood dreams that were like elemental force fields, and having a persistent interest in witchcraft from my early teens.

I remember clearly the first night I ever tried to make something happen. I was subject to the usual teenage frustrations, arising mostly from sex, and used to fly into rages that I found tremendously exciting and exhilarating. I would sometimes take down a few items from the collection of old swords and battle-axes that I was permitted to keep in my room and storm around wielding them at the air and working myself into a beserker rage for the hell of it.

One evening, thus exhilarated, I went out for a walk, still in the afterglow of one of these furies. Some distance ahead of me, I saw a suitable unsuspecting victim in what looked like a bus inspector's uniform. From about thirty yards away, I deliberately loosed a bolt of energy at him by silently pointing and concentrating on his back. Almost immediately he jumped, as if kicked violently in the backside. He turned to regard me with shock and surprise, and I, also shocked and surprised, feigned indifference. Then he covered his embarrassment and went off in another direction. I sincerely hope there were no lasting effects. It was motivated by nothing more than teenage hooliganism and curiosity, and I have rarely used it since. Most importantly, though, I now had proof that there must be something to all those old witchcraft books.

My next discoveries occurred when I experimented with powerful hallucinogens at university. I had expected that these would show me my real soul, the center of the cyclone, as it were. Instead, I saw that in the innermost core of my being there was only the power of will and the power of perception. Everything else was added on and could be stripped away. I began to see that under the apparent order of matter, there was a spontaneous, creative, chaotic, magical force at work. These insights at first appalled me, and it was some years before I accepted and confirmed them.

In the meantime, I took up meditation. At first, I thought that meditation had brought me only calmness, for I had asked little more of it. Later, I was to realize it had brought me a considerable degree of control which was to prove useful in many ways. I began to record my dreams and found that a fantastic world could be opened up with perseverance. Several times, I dreamed of particular things which had just happened to my mother, unbeknownst to me, with complete accuracy.

Encouraged, I started to try to find my hands in dream as a preliminary step to deliberate astral travel. It was then that I encountered the psychic censor. There is some part of one's mind that seems adamant that these experiences are out of bounds. The censor will go to almost any lengths to prevent one experiencing, or remembering that one has experienced, otherworld phenomena. It took me many months to get past this obstacle, but the result was well worth it.

One night, a long-dead relative appeared in a dream. This was enough to shock me into action. Immediately, I found my hands and then pulled them away; suddenly, some part of myself was standing at a place I had arranged to visit, fifteen miles away. The trip was instantaneous and like the sensation of bursting through a balloon. In all subsequent experiences, it has been the same: suddenly, against some huge resistance, I pop through a sort of membrane, and I am allowed to stand and observe the desired place for a few moments before being whisked back. The details of the targets are always exactly as they should be without the usual dreamlike distortions. On the one occasion when I thought I had missed, I suddenly realized that I had arrived upside down at the scene, which was otherwise correct in all details.

About this time, I began to fool around with Austin Spare's death posture and his sigils. I shall not forget the first time I tried to experience an animal atavism through sigils. I had made a sigil to acquire the karma of a cat some days previously. Having concentrated on it for a while and obtained no result, I had more or less forgotten about it on the night that I was walking through a rather poorly lit alleyway in my hometown. I caught a glimpse of something moving to my right and turned to see a huge, fat alley cat sitting on a gatepost. In the moment that our eyes met, something indescribable flashed between us, and suddenly I tore off into the darkness howling and screeching like a feline, completely possessed. Only the lack of grime on my hands afterward convinced me I was not on all fours.

In the flats in the big city where I was living at the time, there was a chap making a progressive descent into madness. Let us call him Ron. Most head doctors would have diagnosed Ron as paranoid schizophrenic. His behavior was bizarre in the extreme. He heard voices continually, and he imagined persecution from the most unlikely sources. One day, he paid me a visit, drawn perhaps by rumors of my odd interests. He was dressed in perhaps five sets of clothes, starving hungry, and almost completely out of his mind. He had been camping out on a heath for some time to avoid the demons in his flat. Having made him as comfortable as possible, I thought I should perhaps try to do anything I could to help.

We went into a room I had prepared for various magical experiments, and I applied the standard procedures of exorcism. Nothing would work. Ron became very defensive and just kept mumbling a stream of dissociated nonsense. Nothing would make him manifest the demons he complained of so that they could be banished. In exasperation, I decided to be his demon. I advanced upon him, snarling and cursing, menacing him with weapons and threats, throwing back at him all the stuff he had been complaining of. For a few minutes, I became his paranoia. Clad in strange robes in a dark room full of burning sulphur, I held his soul at sword point and thoroughly evoked hell all about him.

The effect was remarkable. He opened up and became completely lucid and reasonable, trying to talk his way out of the predicament, using perfect sense and logic with the correct emotional responses. Thereupon, I turned the act off and got us both out of the choking chamber. Ron was then normal for another twenty minutes, during which time we tried to work out how he was going to get his life back together. Toward the end of this, though, he began to slip back into his insane mode, and by the time he suddenly decided to leave, he was completely crazy again.

It was my greatest regret that I didn't have the facilities to detain him and try to do something more for him. As the witch doctors say, a man who gets sick in the head can be helped, but a man with a "bad soul," that is, a long-term head case, often proves intractable. I only saw Ron briefly once again after the psychiatrists had had their way with him, and he seemed a virtual cabbage. Let's hope it was only sedatives.

Around this time, I had my first successes with Spare's death posture. After some months of practice, I suddenly entered a strange ecstatic state which was exhilarating beyond my wildest expectations and seemed to be the gateway to certain magical experiences. I would find myself hovering as a point of awareness outside my body entirely and was allowed to examine the ball of ectoplasmic force that each of us contains in his midsection.

My capacity for tantrums of raging anger sometimes reasserted itself when events became trying. On one occasion, I had lost my door keys in the riotous disorder of my flat and was late for an appointment. I stormed around from room to room, overturning every possible hiding place to no avail, getting more and more angry. Eventually, I stood in the center of the main room and screamed and roared my frustration out. Suddenly, the keys appeared in midair just off my right shoulder and fell to the ground. I swept them up without a thought and hurled myself out the front door. It wasn't until I was fifty yards up the road that the full force of what I had done dawned on my thinking mind. It still tends to give me the creeps.

Shortly thereafter, I set off on a low-budget trip around the world. This proved to be an excellent way of sharpening magical abilities. Long periods of enforced inactivity crossing deserts, and long periods of quiet in the Himalayan mountains seem to bring out a capacity for telepathy. The avoidance of the hazards associated with travel on the cheap is a further stimulus.

If human life is precarious in India, then the life of an Indian feral dog must be about the most demanding and competitive lifestyle possible. As a result, Indian dogs have evolved an intelligence rarely matched by their bloated and pampered relatives in Western countries. You would need to be half psychic to survive as a dog in India, and several that I met were. One in particular would leave whatever he was doing and come to me within a few minutes if I thought intently about him.

There were many Tibetan Buddhist monks in a mountain village I stayed at. There, people are credited with some quite extraordinary abilities. One day, sitting on top of a low building overlooking the market, I decided to see if I could influence one of their number.

Selecting one of the red-robed, shaven-headed monks at random as he entered the market, I commanded him to stop. This he did, despite the fact that he was standing in the middle of the street. "Turn around," I thought. He complied. I let him walk back ten steps the way he had come and then projected the suggestion again. A second time he retraced his steps after turning round again. When he regained his first position, I sent the message a third time. At this he stopped, looked about himself in a confused way, shook his head, and walked determinedly off on his original route. I had lost him. Thinking about it afterward, I suppose it was not really all that clever. Those monks spend all day meditating, and it's not surprising that they are receptive to that sort of thing.

Returning eventually to England, I discovered that autumnal fungi were the subject of much interest. A fellow sorcerer took me out at the beginning of the season and picked me a handful of the aforesaid sacrament, which I ate fresh. About an hour later, lying down in my friend's attic, a magnificent and beautiful vision showed itself to me. It was of a glowing segmented body surmounted by fabulous diaphanous wings. It rotated itself before me so that I could inspect it for some moments and then was gone.

It was not until I went out mushroom hunting a second time that I realized what the vision was. On the second mushroom hunt, I observed that numerous little gnats were buzzing between the mushrooms. When we put some of the mushrooms to dry on a paper, a number of tiny maggots emerged. I had swallowed several of them in the first dose and had somehow picked up on their being. Had I met the lord of the gnats perhaps?

• • •

These are just a few of the easier-to-explain experiences that have come my way during my years of involvement with magic. Many hundreds of instances of telepathy, precognition, coincidence, and spell casting must go unrecorded. I no longer regard these things as strange and unusual. The greatest mystery to me now is why these things are not always accessible and available to us all the time. I consider that anybody who is prepared to strive against his own resistance to these phenomena, "the psychic censor," will achieve results.

CATASTROPHE
THEORY AND MAGIC

C atastrophe theory is a mathematical tool which allows sudden discontinuous changes to be represented by a topological model. Topology is sometimes called "rubber sheet geometry," in which a form can be distorted in any way so long as the basic features remain unaltered. The theory provides a qualitative but not a quantitative description of how a sudden change is likely to occur. It tells us that something unusual is to be expected under certain conditions, but it does not tell us exactly when to expect it. It does, however, tell us how to provoke catastrophes.

The theory is here applied to situations like initiation, illumination, and religious conversion, where there is a sudden change of state. In each case, the theory describes the situation satisfactorily and also throws up much that is unexpected and valuable. It was originally my taste for perfidious wisdom that drew my attention to a theory that most mathematicians still consider "black magic," but having found it to have a certain applicability to magic, I present it herewith.

Catastrophe Theory

I WILL USE MAINLY THE CUSP catastrophe model shown in figure 2. The sheet with the kink in it is the catastrophe surface across which a point representing behavior can move. The Z direction is the behavior axis, and the higher up the axis on the surface it is, the more of a particular behavior it will show. A is the highest point, B is a little lower, D is lower still, and C is the lowest point. The Y axis is a control factor, which tends to move behavior from D to B or from C to A when it is applied. The X axis is a second control factor which causes catastrophes. When there is a large factor of X, any change from A to C or back again will not be a smooth one but will be a sudden change as the point falls over the edge of the fold back down onto another part of the surface. Thus, if the behavior is at point B or D, a change in the amounts of X or Y will cause only smooth continuous changes. However, if the behavior is at point A or C, any large change in Y will result in a catastrophic change as the point falls over the fold.

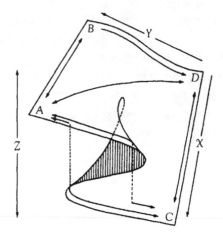

Figure 2. The model for the catastrophe theory.

Applications of the Catastrophe Theory: Occult vs. Materialist Worldviews

This model represents occult versus materialist worldviews. The control factors are Y, the number of magical experiences the subject remembers having perceived; and X, the degree to which the subject is tough minded or rigorous about what he remembers having perceived. See figure 3.

THE TRANSITIONS

B <—> D: an oscillation characteristic of popular occultism, where each bit of uncritically accepted nonsense changes one's mind for the few days that one can remember it.

D —> C: an increasing sophistication of materialist outlook caused by scientific study and rationalization.

B —> A: an increasing sophistication of occult outlook caused by magical study and rationalization.

D —> A: a gradual growth of an occult viewpoint caused by study of magical theories and exposure to magical experiences.

C —> A: a sudden and discontinuous change to occult worldview caused by exposure to magical experiences.

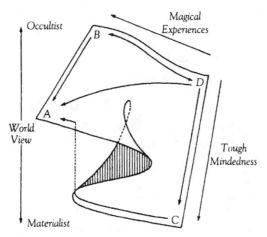

Figure 3. The catastrophe theory applied to occult vs. materialist worldviews. Here the points represent the following: A) tough-minded occultist; B) tough-minded materialist; C) sloppy-minded occultist; and D) sloppy-minded materialist.

Transitions away from A are unlikely, as they would involve eliminating experiences from memory. Any scheme of magical training aims to produce state A in which the subject can be discriminating about occult experience. Some masters insist on a course of mental training that initially forces the pupil into state C so that he can then make the catastrophe revision of worldview to A on being exposed to magical experiences. I tend to regard the route D —> A (a gradual increase in both objectivity and experience) as the easiest route to engineer, although the general purpose of this book is to provoke B —> A transitions.

Applications of the Catastrophe Theory: Change of Belief

We can also use the catastrophe model to show a change from one belief to another. The control factors are Y, emotional commitment to one belief or the other; and X, rationality and irrationality as shown in figure 4. Types of belief (I) and (II) can range from the ideological to the religious and include such beliefs as disbelief.

THE TRANSITIONS

A <—> B and C <—> D: merely represent the weakening and strengthening of belief by the increase or decrease of rationality. This increase of belief with irrationality corresponds to religious revivalism and nationalism.

B <—> D: represents oscillation between rationally held beliefs (I) and (II) on an emotional basis.

A <—> C: represents catastrophic change of irrationally held beliefs (I) <—> (II) caused by change of emotional commitment; for example, religious conversion.

A <—> D and B <—> C: are possible.

A further phenomenon can be shown on the catastrophe surface—that of bifurcation or splitting as shown in figure 5. This figure shows how a descent into irrationality produces two possibilities for strong belief that are very unstable with respect to emotional commitment. This is a technique that has its uses in magic.

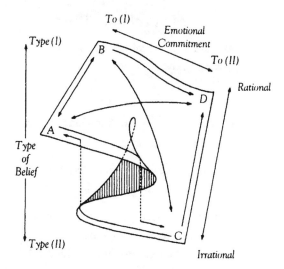

Figure 4. *The catastrophe theory applied to a change from one belief to another. A and C represent strong beliefs that are both emotional and irrational. B and D represent beliefs that are emotional and rational. Note how the element of rationality causes a decrease in the strength of the beliefs.*

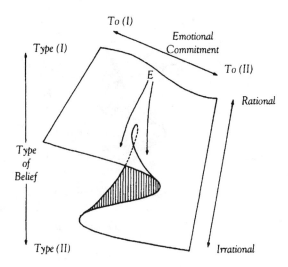

Figure 5. *This is an extrapolation of figure 4 and shows the phenomenon of bifurcation or splitting.*

Applications of the Catastrophe Theory:
Mystical Initiation

The control factors are Y, mystical knowledge; and X, mystical power/ expertise with gnosis as shown in figure 6.

THE TRANSITIONS

D —> A: a gradual increase in both knowledge and power, followed by the better systems.

D —> B —> A: knowledge first, power later; a process which, although safe, may never satisfactorily be completed in time.

D —> C: power first, knowledge later. This type of training or experience often produces heresy or madness. Many mystic masters say that power on its own actually removes the candidate further from the objective of mystical initiation and erects a catastrophe barrier to further progress.

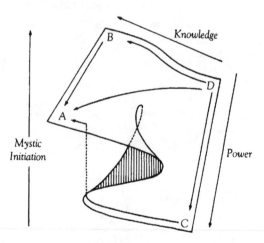

Figure 6. *The catastrophe theory applied to mystical initiation. Here the points represent the following: A) illuminated mystic; B) knowledgeable mystic; C) someone with power but no knowledge; and D) the beginner.*

Applications of the Catastrophe Theory:
Magical Initiation

The control factors are Y, power; and X, knowledge. In a magical initiation, the control factors operate the other way around because the structures of knowledge and power are different, as are their effects on the candidate. It is often simply this difference which leads magicians and mystics to denigrate each other's systems. See figure 7.

THE TRANSITIONS

D —> A: a gradual increase in both knowledge and power.

D —> B: power first, a dangerous path which may lead the candidate into disaster.

D —> C: knowledge first; this type of training usually produces only a dilettante.

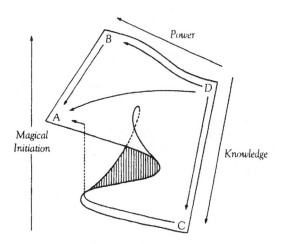

Figure 7. The catastrophe theory applied to magical initiation. Here the points represent: A) illuminated magician; B) powerful, but unwise, magician; C) one who is merely full of ideas; and D) the beginner.

Thus, we see why mystic masters denigrate power for its own sake, and magical masters denigrate knowledge for its own sake. Each removes the candidate further from his objective and necessitates a catastrophic change to achieve the desired effect. Power on its own in magic and knowledge on its own in mysticism do at least bring the candidate some way forward. Both these schemes can be combined in a double catastrophe surface for which I hope the originator of the theory will forgive me. See figure 8.

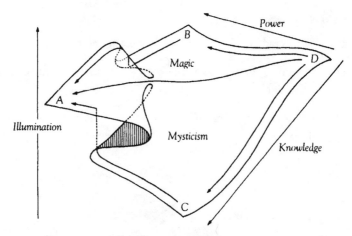

Figure 8. An extrapolation of figure 7—the double catastrophe surface.

Points B and C are on this surface slightly lower than D. Apart from noting how perilously thin the path D —> A becomes at one point, I present it without further comment.

CHAOS THEORY
AND MAGIC

A fter the publication of the first editions of *Liber Null* and *Psychonaut*, chaos theory came to the fore as a scientific discipline. It represents a further development of the principles of catastrophe theory.

Many systems in the world that seemed either random or impossible to predict were found to show extreme sensitivity to initial conditions.

In the classic illustration of the principles of chaos theory, a butterfly changing direction can make the difference between a hurricane occurring or not occurring on the other side of the world months later.

This has let weather forecasters off the hook forever. They can now prove that they cannot predict the weather much in advance without knowing the position and momentum of every air molecule on the planet and very much else besides, including the intentions of every butterfly.

Such extreme sensitivity to initial conditions reflects the way the sorcerer strives to effect large outcomes by well-intuited inputs in advance.

Some chaos theorists believe they have explained at least some of the apparent randomness of the world in causal terms, and they speak of deterministic chaos to imply that only lack of knowledge of initial conditions makes many processes appear chaotic.

However, sensitivity to initial conditions goes right down to the quantum domain where the position and momentum of quanta cannot simultaneously have definite values as well as outcomes which remain truly random. Thus, far from downgrading randomness to mere uncertainty, chaos theory reveals the ladder by which the underlying indeterminacy of the quantum realm climbs easily into the macrocosm of the human- and cosmic-scale events that we observe.

The realization that we inhabit a fundamentally chaotic and random universe leads to profound consequences for the practice of magic. The future and the past remain mutable. Despite that they will often want to do otherwise, magicians need to "enchant long and divine short."

Also in Weiser Classics

Taking Up the Runes: A Complete Guide to Using Runes in Spells, Rituals, Divination, and Magic, by Diana L. Paxson, with new material by the author

Yoga Sutras of Patanjali, by Mukunda Stiles, with a new foreword by Mark Whitwell